Partnering in Design and Construction

Other McGraw-Hill Books of Interest

BARKLEY AND SAYLOR • *Customer-Driven Project Management*
CIVITELLO • *Construction Operations Manual of Policies & Procedures*
CLELAND AND GAREIS • *Global Project Management Handbook*
LEAVITT AND NUNN • *Total Quality Through Project Management*
LEVY • *Project Management in Construction*
O'BRIEN • *CPM in Construction Management*
PALMER, COOMBS AND SMITH • *Construction Accounting & Financial Management*
RITZ • *Total Construction Project Management*
RITZ • *Total Engineering Project Management*
TURNER • *The Handbook of Project-Based Management*

Partnering in Design and Construction

Kneeland A. Godfrey, Jr.

McGraw-Hill

New York San Francisco Washington, D.C. Auckland Bogotá
Caracas Lisbon London Madrid Mexico City Milan
Montreal New Delhi San Juan Singapore
Sydney Tokyo Toronto

Library of Congress Cataloging-in-Publication Data

Godfrey, Kneeland A.
　　Partnering in design and construction / Kneeland A. Godfrey, Jr.
　　　　p.　　cm.
　　Includes index.
　　ISBN 0-07-024038-8 (acid-free paper)
　　　1. Construction industry—Ownership.　　2. Partnership.　　I. Title.
HD9715.A2G62　1995
624'.068—dc20　　　　　　　　　　　　　　　　　　　　　　　　95-245
　　　　　　　　　　　　　　　　　　　　　　　　　　　　　　　　CIP

McGraw-Hill

A Division of The McGraw·Hill Companies

Copyright © 1996 by The McGraw-Hill Companies, Inc. All rights reserved. Printed in the United States of America. Except as permitted under the United States Copyright Act of 1976, no part of this publication may be reproduced or distributed in any form or by any means, or stored in a data base or retrieval system, without the prior written permission of the publisher.

1 2 3 4 5 6 7 8 9 0　　BKP/BKP　　9 0 0 9 8 7 6 5

ISBN 0-07-024038-8

The sponsoring editor for this book was Larry Hager, the editing supervisor was Bernard Onken, and the production supervisor was Donald Schmidt. It was set in Century Schoolbook by Dina John of McGraw-Hill's Professional Book Group composition unit.

Printed and bound by Quebecor/Book Press.

McGraw-Hill books are available at special quantity discounts to use as premiums and sales promotions, or for use in corporate training programs. For more information, please write to the Director of Special Sales, McGraw-Hill, 11 West 19th Street, New York, NY 10011. Or contact your local bookstore.

Information contained in this work has been obtained by The McGraw-Hill Companies, Inc. ("McGraw-Hill") from sources believed to be reliable. However, neither McGraw-Hill nor its authors guarantees the accuracy or completeness of any information published herein and neither McGraw-Hill nor its authors shall be responsible for any errors, omissions, or damages arising out of use of this information. This work is published with the understanding that McGraw-Hill and its authors are supplying information but are not attempting to render engineering or other professional services. If such services are required, the assistance of an appropriate professional should be sought.

This book is printed on recycled, acid-free paper containing a minimum of 50% recycled de-inked fiber.

Contents

Series Preface vii
Acknowledgments ix
Preface xi

Part 1 Single-Project Partnering

Chapter 1. Introduction 3

Chapter 2. How Public-Sector (Fixed-Price) Partnering Got Started 19

Chapter 3. A Contractor's Partnering Baptism 31

Chapter 4. How One Contractor Switched Styles from Confrontational to Partnering 53

Chapter 5. Can Architects Return to the Construction Site via Partnering? 65

Chapter 6. One Partnering Success Secret: Set High Goals 75

Chapter 7. The Partnering Process—A Project Management Strategy to Improve Quality and Field Productivity 89

Part 2 Other Project Partnering Applications and Implications

Chapter 8. Case Studies in Strategic Alliances 113

Chapter 9. Your Lawyer as a Project Team Member 127

Chapter 10. Alternative Dispute Resolution as a Partnering Tool 137

Part 3 Nonproject Uses of Partnering

Chapter 11. Steel Erectors and OSHA Partner to Dramatically Improve Safety — 165

Chapter 12. How One Contractor Partners with Its Employees — 179

Chapter 13. How One Design Firm Partners with Foreign Coworkers — 201

Chapter 14. How Granite Rock Co. Won the Malcolm Baldrige National Quality Award — 211

Chapter 15. Why One Contractor Calls Partnering "Project Quality Planning" — 231

Chapter 16. How Partnering and Empowerment Made an MBE Effort a Winner — 245

Chapter 17. How the New Model Subcontract Was Created — 257

Part 4 Tomorrow

Chapter 18. Can Partnering Transform Design Practice? — 269

Chapter 19. The Future of Partnering — 283

Index 299

Construction Series Preface

Construction is America's largest manufacturing industry. Ahead of automotive and chemicals, construction represents 14% of this country's gross national product. Yet it is unique in that it is the only manufacturing industry where the factory goes out to the point of sale. Every end product has a life of its own and is different from all others, although component parts may be mass produced, or modular.

Because of this uniqueness, the construction industry needs and deserves a literature of its own, beyond reworked civil engineering texts and trade publication articles.

Whether the topic is management methods, business briefings, or field technology, it will be covered professionally and progressively in these volumes. The working contractor aspires to deliver to the owner a superior product, ahead of schedule, under budget, and out of court. This series, written by constructors, for constructors, is dedicated to that goal.

M. D. Morris, PE
Series Editor

Acknowledgments

This book is the work of many hands.

When M. D. Morris of McGraw-Hill suggested that I write a book on partnering, I told him, "I can't—too busy." To which he replied, "You don't, you Tom Sawyer it." As Tom did when charged by Aunt Polly with whitewashing the fence, you persuade a dozen or so other people that it will be an honor to write a chapter. Thanks, Dan.

Eighteen partnering leaders kindly responded to an invitation to write a chapter. Their readers are in for a treat. The authors show that partnering is not limited to single projects alone, and their chapters show that partnering is key to successful A/E/C firms, professionals, and people.

I must thank Larry Hager of McGraw-Hill for his encouragement (and an occasional push when the project was idling), his responsiveness when chapter authors posed perplexing questions, and much else. If the book is well received, it will be due in no small measure to Larry's partnering support.

At my employer, the Institute of Management & Administration (IOMA), John Marqusee, David Foster, and Lee Rath have provided the most supportive environment imaginable. When I proposed four years ago that we start a contractor business management newsletter, they said yes—and had IOMA's telemarketing department call 40,000 contractors and invite them to subscribe. It was in *Contractor's Business Management Report*'s first two years that I learned about the construction partnering movement. IOMA's generous support of my travel, subscriptions to periodicals, and phone interviews has been vital to a newsletter that has, I think, a growing reputation. Thanks for putting up with an editor sometimes preoccupied with nonnewsletter business.

My wife Anne has been not only cheerleader but home-office manager. She remembers and sees things I do not. She supports when that is what is needed, and pricks the balloon of self-importance when necessary.

She shows every day, mostly by her actions, that the key to partnering is service. This book is dedicated to you, Anne.

Preface

Partnering is perhaps the hottest trend in construction today. It should be discussed in depth in a book, and it has—two books on the subject have been published recently. This one is different:

1. It is not primarily a "how to" book. The tools used in single-project partnering are by now widely known, but not how and why partnering was born. Larry Bonine (Chap. 2) recalls.
2. Use of partnering is growing fast, but there is a danger that it will be merely a passing fad. User feedback, systematically recorded, is needed. Lou Bainbridge (Chap. 7) reports lessons learned from the first national survey of participants in partnered jobs.
3. Case histories: Michael Murphy (Chap. 3) recounts how partnering transformed his company and his business approach. Clement Mitchell (Chap. 4) and the team of Jim Gans, Gretchen Gagel McComb, and Ed Wambsganss (Chap. 6) describe their first experiences with partnering.
4. Successful partnering eliminates claims and lawsuits, but even committed partners sometimes disagree and even fight. Attorney Fred Lyon (Chap. 9) suggests how to use your lawyer to (a) support partnering, not fight it, and (b) write contracts that accommodate partnering. Attorneys Robbie MacPherson and Richard Steen (Chap. 10) describe the options in alternative dispute resolution, and illustrate their use with case histories.
5. Single-project partnering on public jobs sprang from multiproject strategic partnering with private clients. Anthony Costonis (Chap. 8) shows how three leading industrial owners use multiproject partnering with their contractors.
6. Rates of construction worker injury and death are plummeting. At the heart of this safety revolution is contractor–worker partner-

ing. Construction safety consultant Steve Miller, a veteran iron worker and now a leader in safety reform, tells how Colorado steel erectors and OSHA are working together (Chap. 11).

7. Contractors who honor their workers—partner with them—are repaid in kind. Nick Bouler (Chap. 12) says his construction company is founded on a commitment to employees, and that is a key reason why it has grown to be one of the nation's larger and most respected firms.

8. The same is true of design firms. Henry Michel (Chap. 13) tells how partnering with his firm's foreign engineers—that is, acting just opposite those who exploit low-labor-cost engineers from developing countries—has sparked its growth to 500 foreign employees overseas. And Jim Bradburn (Chap. 5) explains his vision of using partnering as a tool for bringing architects back on site during construction.

9. As partnering is to the project, total quality management (TQM) is to the company. Leading firms are internalizing partnering lessons, applying them systematically to additional projects. TQM is not mysterious and not as irrelevant to the architecture/engineering/construction (A/E/C) industry as some have thought. It is simply a systematic approach to honoring both customers and employees—that is, partnering with them. Representatives of two leaders in contractor partnering and in using TQM, Ron Deffenbaugh (Chap. 15) and Bruce Woolpert (Chap. 14), tell how.

10. General contractor/subcontractor partnering can also pay off in a big way. Gerry Graff, 1994/95 president of the American Subcontractors Association, tells how general contractors and subcontractors cooperated in writing a new model building subcontract (Chap. 17). And former contractor Lorry Bannes (Chap. 16) relates how his team partnered with minority subcontractors in St. Louis, not only creating a good building but also helping some of the subs grow into successful, long-term businesses.

11. Finally, we include two views on where partnering is leading A/E/C firms, by consultant Kyle Davy (Chap. 18) and partnering facilitator Jerry Pitzrick (Chap. 19).

The concepts of partnering are being applied in many diverse ways. This book reports on a remarkable number of them.

Kneeland A. Godfrey, Jr.

Contributors

Louis R. (Lou) Bainbridge heads the quality and productivity improvement group at contractor management consultant FMI in Denver. The national leader in construction project partnering, Lou's group has more than 425 partnering efforts underway, ranging from $400,000 to $800 million. Their combined construction value is $15 billion.

With FMI since 1981, Lou spends 75% of his time on the road nationwide. He's been on the FMI board since 1987 and is certified by the Institute of Management Consultants. Lou has taken the lead in educating industry groups in successful implementation of single-project partnering.

Lorry T. Bannes, P.E., is president of Bannes Consulting Group (St. Louis), where he's a construction arbitrator and management consultant, and does construction planning and design-phase consulting.

After receiving the B.S. in civil engineering from St. Louis University in 1957, Mr. Bannes was a structural engineer with a St. Louis firm. He then spent 12 years with Gamble Construction Co. in that city, rising from construction engineer to president and becoming the first non-family member in that post. In 1972 he and a coworker resigned and formed Bannes-Shaughnessy Inc. (St. Louis), a construction firm, where he stayed until 1989, when he formed his present company.

xiv   Contributers

Larry S. Bonine is Director (staff chief) of Arizona's Department of Transportation, whose partnering operation is one of the largest among U.S. public agencies. He came to Arizona in 1993, after serving with the Bechtel/Parsons Brinckerhoff joint venture as Project Wide Area Construction Manager for Boston's Central Artery/Tunnel Project.

Earlier he spent 20 years with the Army Corps of Engineers, and was District Engineer (chief) of both the Mobile, AL, and Little Rock, AK, districts. In the American Association of State Highway and Transportation Officials, Larry chairs the Standing Committee on Quality.

Nicholas O. Bouler, III, is manager of Construction Personnel & Training for BE&K Construction Co. (Birmingham, AL), where he has overall responsibility for field hourly personnel issues and craft training, nationwide, for the several-thousand-worker contractor.

Nick is an attorney, and previously served in BE&K's corporate legal department. Earlier he was a partner in a small legal firm, specializing in trial and administrative practice with an emphasis on employment law. He is a frequent speaker and panelist on employment-related issues before business and professional groups.

James Henry Bradburn, AIA, is principal in charge of design technology and operations at the architectural firm C.W. Fentress J.H. Bradburn and Associates (Denver).

For 12 years a project architect with Kevin Roche John Dinkeloo and Associates, he has been called an architect with the heart of an engineer. He seeks to maintain the highest levels of performance in production and architectural technology, and has won several awards for technical achievement. Jim has lectured on project partnering and team building, and is on the board of the Colorado Christian Home, a treatment center for abused and neglected children and their families.

Contributers   xv

Anthony F. Costonis, PhD, is founder and president of Corporate Development Services (Lynnfield, MA), a contractor management consultant. Dr. Costonis grew up in a construction family and continues in the business—over the past quarter century, CDS has worked with hundreds of contractors nationwide and worldwide.

He has lectured on managing construction businesses with ABC's Executive Management Academy, AGC's Advanced Management Training programs and the National Electrical Contractors Association's Executive Study Program. He wrote *Planning for Growth and Profit*, the first of ABC's executive-level management books.

Kyle V. Davy, AIA, is a management consultant in Berkeley, CA, who works with design firms seeking to optimize their operational performance through quality improvement programs, project management, and practice-management innovations.

He leads the operations faculty at the Advanced Management Institute for Architecture and Engineering (San Francisco) and teaches its courses in project management, quality management, and practice management.

Earlier he was COO for a 45-person architectural firm and designed numerous projects. A graduate architect, he also has an MBA from Stanford University.

Ronald L. Deffenbaugh is vice president of total quality management at the building contractor McDevitt Street Bovis (Charlotte, NC). He works with executive management in introducing the quality principles of customer satisfaction, continuous improvement, management by fact, and respect for people.

Prior to coming to Charlotte in 1992, Ron was quality manager for three years and a senior estimator for four years in the firm's Washington, DC, division. Under his leadership the division began its pursuit of the U.S. Senate Productivity Award for the state of Maryland, which it has since won in 1994.

Key to winning the award was the firm's Project Quality Planning process discussed in Ron's chapter, which he helped pioneer. He earned his B.S. degree in construction technology from Bowling Green (Ohio) State University.

E. James Gans is chief of the Clark County (Las Vegas) Sanitation District, the state's largest. While in that position over the past 15 years, he has also directed the county's Salinity Control and Waste-Flow Reduction programs, been a member of the U.S. Colorado River Floodway Task Force and chaired the national convention of the then-Water Pollution Control Federation.

For four years Jim studied civil engineering at the University of Nevada, Las Vegas, but he was forced to change majors because the school did not then offer a degree in that field. So he got the BS in psychology, "not realizing how valuable that subject area would be."

Kneeland A. (Ned) Godfrey, Jr. edits three monthly A/E/C-firm management newsletters for the publisher IOMA (New York City): *Contractor's Business Management Report, Design Firm Management and Administration Report,* and *Principal's Report.*

A civil engineering and journalism graduate, he edited ASCE's magazine *Civil Engineering* for 15 years. Still earlier he was public information officer for the American Association for the Advancement of Science, and for Northwestern University. His first job was as an editor for *Rock Products* magazine. He is a recent president of the Construction Writers Association.

In nearly 40 years of semitechnical writing, he's sought to make it both technically sound and interesting to readers other than specialists.

Gerry Graff, president of Graff Flooring Contractor Inc. (Albuquerque, NM), was 1994–95 president of the American Subcontractors Association (Alexandria, VA).

A veteran of over 20 years in a family-owned construction business (his state's oldest commercial flooring operation), Gerry has for 14 years been president.

He was one of the steady hands who kept on track the intersociety task force that crafted the new model building subcontract which he describes in his chapter.

V. Frederick Lyon is a partner in the national law firm of Lyon and McManus, with offices in Washington, DC, Orlando, and Atlanta. He specializes in construction law and focuses on energy and electric power-related issues. The firm represents both electric utilities and major contractors building powerplants.

Fred served a judicial clerkship in the U.S. Court of Claims and then entered private practice, in 1977. He is a charter member of the American College of Construction Lawyers. A frequent lecturer and author on risk control, he has a national reputation in innovative dispute avoidance and resolution, with an emphasis on teambuilding and contract preparation.

Robert J. MacPherson is a partner in the law firm of Poster & Rubin, which has offices in New York City and Maplewood, NJ. Since entering practice in 1980, he's focused on construction, primarily prevention and resolution of disputes.

Robbie, as he likes to be called, has arbitrated several complex matters, representing suppliers, subcontractors, general contractors and owners. He is coauthor of the book *New York Construction Law Manual*, published by Shepherd's/McGraw-Hill.

The 1994–95 chair of the Dispute Resolution Section, New Jersey State Bar Association, he is also a member of the Supreme Court of New Jersey's Committee on Complementary Dispute Resolution.

Gretchen Gagel McComb is a consultant with the quality and productivity improvement group of contractor management consultant FMI (Denver). She assists contractors in assessing their management practices, and in establishing effective project-control systems. She also facilitates design and construction project-partnering efforts.

Previously, Gretchen was an operations manager with Ralston Purina Co. and Coca-Cola, responsible for implementing TQM in union and nonunion facilities employing up to 1,000 personnel. The results were significant gains in productivity, measurable cost savings, improved employee morale and a team-oriented environment.

She grew up in a family which owned a construction equipment dealership, and her family includes former national leaders of that industry's trade group, the Associated Equipment Distributors.

Holder of a mechanical engineering BS and an MBA, Gretchen has taught public speaking at the university level.

Henry L. Michel, P.E., is chairman of Parsons Brinckerhoff Inc. (New York City), one of the largest U.S. engineering firms.

Prior to joining the firm in 1965 he spent 15 years as head of a European-based engineering firm which had such design credits as Baghdad University in Iraq and the World Health Organization headquarters complex in Geneva, Switzerland. That richness of foreign experience (and the many contacts made) are among the reasons PB has built up one of the largest international design practices among U.S. firms.

Henry has been the firm's principal in charge of the $2 billion Metropolitan Atlanta Rapid Transit Authority rail system in the 1970s, and in the past five years, the $16 billion Taipei Mass Rapid Transit system project in Taiwan.

He has led the International Road Federation, and, back home, ASCE's Civil Engineering Research Foundation.

Steven P. Miller, a national leader in steel-erection safety, in 1990 founded Miller Safety Consulting (Thornton, CO).

Steve is a 25 year construction veteran, with focus on managing craning and rigging work, and safety management. Employers have included NASA (White Sands, NM), and contractors Martin K. Eby Co., Peter Kiewit and Sons, and Research Cottrell. For six years he worked with steel erection and rigging contractor LPR Construction (Denver), before starting his firm.

Steve has also:
- Taught thousands of safety and management hours on HazCom, trenching, rigging, craning, climbing, and ironworker safety;
- Developed a safety-training video for local contractors;
- Helped implement one of the first, 100% fall protection (6 foot fall rule) programs at a national contractor;
- Developed and patented an ironworkers' and riggers' safety device (see illustration in his chapter) that's gaining wide national usage;
- And serves on the OSHA Advisory Subcommittee on Fall Protection, under the parent "negotiated rulemaking" committee which is developing OSHA's new Steel Erection Standard.

Clement V. Mitchell is president and CEO of MCI Constructors (Woodbridge, VA), a builder specializing in mechanical-intense projects such as wastewater and heating plants.

A 23-year construction veteran, Clem has been with MCI for 15 years in such positions as project manager, purchasing agent, and vice president. He formerly held management positions with Bechtel and Catalytic Corporation.

MCI Constructors is the primary subsidiary of FICON Corp., a subsidiary of Fischbach Corp., one of the largest U.S. contractors. Clem is also president and CEO of FICON.

Michael B. Murphy is managing director of Donald B. Murphy Contractors, Inc. [Federal Way (Seattle), WA], a specialist in geotechnical construction, specialized steel rigging, and heavy rigging.

After earning a magna cum laude BA degree in 1972, Michael went to work as a construction laborer, moved up to operating engineer, foreman, project superintendent, and then project manager. He has worked jobs from the Florida Keys to Alaska's North Slope, and on the other side of the Pacific Rim.

Jerry Pitztick is a graduate civil engineer who's been with the large building contractor Mortenson Co. (Minneapolis) for his entire, 21-year career. He has served the company in an impressive variety of roles. These have included project engineer, superintendent, project manager, division manager, director of technical services, and quality director. He has been involved with commercial building projects, hospitals, hotels, and electric power plants.

Since 1988 he has served as facilitator of total quality management and partnering efforts with Mortenson and others. That is, as facilitator at partnering retreats on some 30 non-Mortenson jobs, and in that capacity at nearly 50 Mortenson building projects.

Richard H. Steen is vice president of Hill International Inc. (Willingboro, NJ), where he concentrates on resolution of complex construction and environmental disputes through negotiation, litigation, or ADR techniques.

An experienced arbitrator and mediator, Rick has chaired the Section on Dispute Resolution, New Jersey State Bar Association. He's a member of the Standing Committee on Complemenatry Dispute Resolution, New Jersey Supreme Court.

He has lectured for several groups and has written numerous articles. He's served as Legislative Counsel to the New Jersey State Bar Association and Staff Counsel to Republican members of the New Jersey General Assembly. A graduate of Seton Hall University School of Law, he was a founding director of the university's Legislative Bureau and the first editor of the *Seton Hall Legislative Journal*.

Edwin F. Wambsganss, a graduate civil engineer, is president and CEO of the civil and process contractor Western Summit (Denver). The firm builds power-related facilities, pipelines, bridges, highways, municipal sewer and water projects, medical facilities, and environmental monitoring and ground water contamination reclamation projects.

In 1973 Ed founded Western Empire Constructors. He later started the merit shop contractor Summit Constructors, and eleven years later combined the two.

Bruce W. Woolpert, co-president and CEO of Granite Rock Co. (Watersonville, CA), has a biographical information sheet with interesting juxtapositions:
- Just below "Co-membership chair, U.S. Quality Council, Western U.S. Chapter," is "Trustee, Foundation for Aggregate Research and Technology, University of Texas."
- And just below "Member of Executive Committee" of Granite Rock appears "Chair, Serious Accident Review Committee" of the firm.

These juxtapositions help clarify why Granite Rock was the first construction-industry firm to win the Malcolm Baldrige National Quality Award. Granite Rock is in the aggregates business and has used quality tools to jump ahead of its competition. One way they've done it is by achieving superiority in safety—which in turn was achieved by the top people making safety a, or *the*, top management priority.

Part 1

Single-Project Partnering

Chapter 1

Introduction

Kneeland A. Godfrey, Jr.

The star attraction at the Associated General Contractors' (AGC) 1992 Dallas Convention was a speech by President George Bush. It was an impressive show: the 1000-person audience was required, for security reasons, to be seated two hours beforehand. There was a color guard and lots of flags.

At another session of the same convention, an audience perhaps one-fifth as large heard a presentation by less well-known people, but the topic, partnering in design and construction, was probably far more important to those in the architect/engineer/contractor (A-E-C) business. Arguably, partnering is the hottest construction industry trend of the 1990s.

At the session on partnering, there were no flags or brass ensemble, but there was something equally striking, if quietly so: Throughout the two-hour session the three speakers stood up, side by side, before their audience. Afterward I asked session leader David Roberts, an engineer for the Dallas suburb of Garland, why they did this, and he said, "to symbolize that we're a team." Roberts spoke first, but when he hesitated, or forgot a key point, one of the others immediately spoke up. To reinforce for their audience the idea of what partnering is, the three A/E/C experts were partnering in their AGC presentation.

The case-history subject was design and construction of a $10 million addition to Garland's Rowlett Creek wastewater treatment plant. The two other speakers were the designer's point man, project manager Ron Sieger of CH2M Hill (Dallas), and the general contractor's project manager, Lynn White of Martin K. Eby Construction Co. (Fort

Worth). Despite major setbacks, thanks to partnering the job came in on schedule and within budget, and the designer and the contractor both made money. As a result, the two firms began marketing their partnering expertise.

What is Partnering?

Partnering can refer to many different relationships, including:

- Single-project partnering
- Multiproject, strategic partnering between a contractor and a client
- Use by a contractor of the same partnering process over many projects, as a key building block in a total quality management (TQM) or other quality program
- Contractor–employee partnering in the pursuit of safety goals
- Partnering to bring women and minority workers into the employee mainstream, in a way in which they are not coddled, excluded, or exploited.

Some remarkable successes in the use of these ways of partnering are told in this book, in part by case histories.

Single-Project Partnering with Private Clients

Partnering between contractors and private clients is as old as construction itself. The Cincinnati construction firm Al. Neyer has passed the age of 100, and President Don Neyer says that the reason is a "client-first" attitude. The company rectifies work clients dislike and provides a two-year warranty, twice the area standard. When clients call Neyer, asking if they should rent a building or construct their own, Neyer often advises renting. "If you serve your clients' best interests, you'll create a friend. They'll come back to you when they need to build," he told *Building Business,* the newsletter of CPA firm Grant Thornton (Chicago).

Charlie Boyd, a manager with the large, 100-plus-year-old general contractor Sundt Corp. (Tucson), told me, "Sundt wouldn't have grown this big or lasted this long if we hadn't partnered with clients and subcontractors." Subcontractors are honored, not exploited, so the best want to work for Sundt.

When Stuart Dobson, vice president of general contractor Whiting-Turner (Baltimore), spoke at an American Subcontractors Association meeting, he was asked: "What's the difference between a partnered job and a well-run one with no formal partnering?" He replied, "Nothing."

Dobson recalled, "When I was project manager on bridge projects in the late 1970s, we always had a preproject meeting involving us as general contractor, the owner, and key subcontractors. As a result, quality was up and accidents down." (As this book will make clear, Dobson is not quite right in saying that formal partnering involves nothing new, but the spirit on not a few informally partnered jobs is just as positive. That spirit is perhaps success ingredient number one.)

Single-Project Partnering on Public Projects

In Chap. 2, Larry Bonine, then of the Corps of Engineers' Mobile (Alabama) District, recalls how it started back in 1988: "As District Engineer [the boss], it was my job to get new business and foster customer relations. Our most sensitive customer was the U.S. Air Force....They were a captive customer and unfortunately we behaved accordingly....We believed the Air Force could have made a strong case for building future projects in-house. Treating the Air Force as a captive customer would no longer work" for the Corps.

In private construction, the Construction Industry Institute (Austin, Texas) was then documenting that multiproject strategic partnering was getting jobs done faster, cheaper, and with fewer injuries. But each federal job had to be bid singly—there were no multiproject deals (with some exceptions). What could the Mobile District do?

Bonine was invited to write a chapter in this book because he has been called the father of single-project partnering, although Chuck Cowan usually gets deserved credit for being its chief apostle. As Bonine writes here, credit should go also to Dan Burns, then Mobile District construction chief.

Burns, now with the Corps' headquarters in Washington, DC, recalls: "The idea was adapted from several sources, first of all the Construction Industry Institute's multiproject strategic partnering research in the private sector. The second source was Donald Mosley and Carl Moore at the University of South Alabama, Mobile, who had done nonconstruction team building.

"Why team building? Because engineers know how to manipulate materials, not people. So I looked for industrial psychologists and organizational effectiveness people, and found Mosley and Moore." [Editor's note: See the "Pre-Workshop Survey," that Mosley and Moore supplied, Fig. 2.1. This "people tool" helped the project team get a handle on attitudes during that first (1988) public-project partnering retreat.]

Burns recalls: "We wanted to clear the [emotional, human-relations] decks so project problems could be addressed." If the Corps could transform each person from an individual to a member of a team, it stood to benefit from

> ...better answers, less defensiveness. Because several brains would be attacking a problem. In contrast, answers developed by one person tend to get compromised in the process of selling them to others.
>
> "Example? Shop drawing submittal. In those days, when the engineer decided a drawing was unacceptable, he'd sometimes write, 'Doesn't Conform,' and return it. No information as to how or why it failed. The contractor had to guess again and resubmit. And often a drawing would cross many desks at each organization—long, slow process. Often the contractor would conclude, 'I'll do it the way I'm told, but it's not what the contract calls for.' Then the contractor would have the basis for a claim for an extra.
>
> "Instead, if we could partner, we could throw both parties' perceptions on the table, we might get better solutions, quicker, and our people would become better managers.
>
> "And it seems to be working. In 1986 the Corps nationwide had 1103 claims with total value $366 million, and thanks in part to partnering, by 1993 the numbers were 532 claims and $228 million."

Burns remembered two other gems:

> 1. Partnering can make work more satisfying. Before, if there was a problem, the person would say, "I've got a problem—got to solve it myself." Today, if anyone on the team has a problem, it's everyone's problem. As a result, work is more satisfying on partnered jobs. Don't discount the value of this satisfaction.
>
> 2. Partnering facilitators need not be construction experts. I recently found an industrial psychologist at a northeast Washington state college, briefed him on partnering, he facilitated the pre-project retreat, and with considerable success.
>
> With experience on several jobs, a facilitator no doubt will improve. But I think many industrial psychologists would do fine. You don't necessarily need to limit yourself [to those offering themselves as facilitators].

Case Histories

1. *A $4 million tieback contract on the Bonneville Lock project on the Columbia River is among the smaller jobs on which partnering has been applied.* But I think Michael Murphy's chapter (Chap. 3) on subcontractor Donald B. Murphy Contractor (Federal Way, Washington) is among this book's most powerful. Murphy says that partnering on this job not only saved his company, but changed his attitude toward work and business. Read his chapter carefully, with an open heart. You will find that his is open to you—he's a brave man. He also has a sense of humor.

Murphy tells why having fun on the job is one of the benefits of partnering:

> Are you working more and enjoying it less?...Resolving disputes requires time and energy....If the answer is "yes," then why not try and do something about the situation?...If possible, why not introduce an element of fun, or at a minimum reduce the stress associated with the confrontation related to disputes. Why not solve problems immediately, rather than let them fester and grow into disputes?
>
> Murphy's firm lost money on the Bonneville contract, but since then its partnering attitude has paid off at the bottom line, too. "Successful negotiations and early payment of equitable adjustments" has improved the firm's liquidity 50 percent. Sales have risen 40 percent, and margins are up 10 percent despite increased competition.

2. *How a contractor once labeled a "claims contractor" was converted to partnering.* Chapter 4 is the story of MCI Construction (Woodbridge, Virginia), which, after making dozens of claims on a Corps of Engineers military job, was terminated for convenience by the owner. (MCI makes a good case that the villain was the Corps and not it.)

Then MCI President Clement Mitchell met the Corps' Col. Richard Sliwoski, the two hit it off, and MCI's business approach to federal contracts was transformed from confrontational to partnering. The result was that, on a Ft. Dix (New Jersey) wastewater project:

> The participants agreed to work together as a cohesive team to produce a quality project, in accordance with the contract, on time, within budget, and safely, while enabling the contractor to earn a fair profit. Members of the partnering team agreed to deal with each other in a fair, open, trusting, and professional manner. In that spirit we agreed to communicate openly, resolve problems, and make decisions at the lowest possible level. We agreed to...talk before we write, and promote pride in workmanship....We maintained these commitments by having periodic feedback sessions to determine whether we were meeting our objectives. We were sensitive to each party's problems. A meeting was never adjourned until everyone was heard, and the group committed to resolve any problems. The enemy had become the problem, not each other.

3. *How can you get the architect back into the field, and thus benefit the project?* Use partnering. James Bradburn, partner in the Denver architectural firm C.W. Fentress J.W. Bradburn, concedes in Chap. 5 that every set of plans and specs is incomplete and has errors. When the contractor in the field has questions about them, the architect should be there to interpret. Not only is a project hurt by the architect's absence from the field, so are architects:

Architects are trained to understand how things are built, how to investigate the serviceability of materials and products, and how to manage and work with people to achieve mutually satisfactory results. When the architect is removed from the construction process, this training is not utilized. Consequently, the skills are diminished or lost.

Bradburn is encouraged to believe that, thanks in part to the coming of partnering and design/build, we are returning to the centuries-old, master-builder approach.

How Can Partnering Help a Project?

Partnering and project schedules

Partnering seems to help the project schedule more than it helps any other project measure. For example, on a recent, $9.6 million city sewage-plant upgrade in the Dallas suburb of Garland, Texas, the schedule called for completion in 18 months, but general contractor Martin K. Eby Construction Co. (Fort Worth) finished the job in 15 months—a 16 percent time savings.

Impressive enough—until you learn that shortly after work began, the job was hit by an 8-in rainstorm that added 4 months to the schedule. The near-record rain flooded the site, leaving several inches of river-bottom silt behind. Because Garland's wastewater plant-operating experience had familiarized it with vacuum equipment, the silt was removed relatively quickly: The city rented the equipment, manned the vacuum trucks, and *did not charge the contractor.*

Soon contractor Eby was able to return the favor. The heavy rain so softened the site where Eby would have to place 70,000 yd^3 of fill that the owner's staff was not able to maintain the site's haul roads, spread and compact the fill, and accept incoming trucks. Eby volunteered a bulldozer and operator, *and paid for both,* since the city had no budget for this.

The early completion was possible largely because the three key parties (contractor Eby, owner Garland, and engineer CH2 M Hill of Dallas) had become teammates. Despite the setbacks, Eby made more money on the job than they had expected.

Partnering and construction safety

Of all aspects of project management, parties agree that partnering helps safety more than it does any other. For instance, consider the construction of a sewage-treatment plant addition at Clark County, Nevada (Chap. 6). The owner-contractor-engineer team set extremely aggressive safety goals—and beat them: 300,000 labor hours of field labor with zero lost-time accidents. The key, as always, was the leadership of the general contractor, Western Summit (Denver), and its president, Ed Wambsganss.

Project budget control

In late 1993, results of the first major national survey of partnering's impact, on over 100 projects, were reported by Lou Bainbridge of FMI (Denver). As Bainbridge reports in Chap. 7, the survey puts some numbers to users' opinions of partnering. For example, on project budget control, some 3.3 times as many respondents said partnering helps, as said it hurts.

Why Are Decisions Often Better on Partnered Jobs?

Two keys to success are coworker empowerment and structured issue escalation.

Empowerment

When a partnered project runs particularly well, often one of the reasons is empowerment, which is defined by Tom Warne in his book, *Partnering for Success* (American Society of Civil Engineers, New York, 1994), as follows:

> Empowerment is the delegation of authority and responsibility to the lowest possible level in an organization. It is not an abrogation of responsibility, but a transfer of authority and accountability to individuals...closest to the problem [and] best equipped to make...decisions.

Of course, there are limits to how far down the organization chart you can delegate. Warne gives an example:

> In every state there is a statute that details which engineering drawings must be signed and sealed by a professional registrant....It would be improper and illegal to attempt to empower a draftsman to sign and seal drawings.

What if a delegatee makes the wrong decision? This is key. Warne writes:

> True there may be times when [in an empowered organization an individual] makes a decision a senior manager does not fully agree with. On these occasions the...temptation will be for that manager to change or override the subordinate's decision. This reaction should be avoided at all costs unless someone's life or property is in jeopardy.

A questionable decision by a subordinate presents the manager with an opportunity to train the subordinate so he fully understands what the course of action should have been. However, great care should be taken to prevent the subordinate from feeling unsure about making future decisions. Nothing will discourage empowered employees more than being "slam dunked" by a manager after making a particular decision.

Warne notes that, prior to partnering, Arizona Department of Transportation (ADOT) resident engineers were empowered on their own to make spending decisions up to $15,000, and district engineers, up to $50,000. Those numbers have been raised to $50,000 and $200,000, respectively. Today upper management needs to spend less time reviewing decisions made by others.

Issue escalation ladder

As Warne notes, a second key to partnered jobs is the "issue-escalation ladder." Disagreements between the parties are inevitable, on partnered jobs as on any others. If disputes are resolved quickly, the project moves ahead smoothly. If not, like an untreated wound that may fester and grow worse, any problem can grow into a claim.

Thus, decisions resolving disagreements among the general contractor, subcontractors, architect/engineer, and owner must be resolved expeditiously. Agreement on how many days an unresolved issue can sit on someone's desk is part of preproject partnering agreements on ADOT jobs.

Warne spells out four "musts" in making issue escalation work:

1. Issues must be escalated at the same time by both organizations in dispute. No one can be allowed to go behind another's back.

2. No one can be allowed to sit on a problem and prevent a solution. "No decision" is unacceptable.

3. "We must change the paradigm about admitting to our supervisor that we cannot solve a problem. [Warne's] experience has shown that employees are at first reluctant to escalate problems because of this paradigm." Managers need to coach and encourage their people to make decisions rapidly.

4. Be reassured that empowerment and "the ladder" do not push all decisions to the top of the hierarchy. After 100 partnered projects in two years, Warne reports, only about a dozen issues have been escalated to the top of the ADOT ladder. Top-management time is not wasted—they are not forced to become "superproject managers."

Who Should Be Invited to Participate in Project Partnering?

In 1993, the first year in which the AGC awarded its Marvin M. Black Excellence in Partnering Awards, eight contractors won awards. Each was asked: "How can partnering be improved on your next project?" The almost universal reply was: "Be more inclusive." That is:

- Do not start partnering at the beginning of construction, but at a project's head end, the start of design.
- Invite key subcontractors into the process.
- Reach deeper into your own ranks; invite key superintendents and some foremen.
- Invite your rank-and-file coworkers.
- Fold into the process parties not yet on board at the time of the preproject partnering retreat.

Do you see a pattern? The feeling of teamwork, the spirit of "we are special," is not going to be lost if you invite too many parties in.

Still, many people believe that the spirit will be lost. Why? On the surface, they want to limit participation in the name of efficiency. Dig deeper, however, and you will find that they think if they spread the gift of partnering too broadly, they will lose it. Partnering, however, is itself partly spirit—and spirit is not a commodity of which there is a limited quantity.

Of course, on a large job it would be impractical or even impossible to invite all the workers to a preproject retreat. If this is the case, find other ways to include them. For example, as each subcontractor comes on the job, gather all their workers together and introduce the project partnering charter.

Partnering Is a Matter of Spirit

Does partnering work because those who have done it have been the best contractors anyway? Charles Cowan, former head of the ADOT, says that many people assumed this was the case. He reports that the ADOT once categorized contractors into groups according to their claims history, but that all have stopped making claims since partnering was adopted.

In his foreword to Warne's book, Larry Bonine, Cowan's successor, writes that on an ADOT partnered job that is working right, "Somehow the...atmosphere is different—more highly charged and exciting. People talk to each other and there is an air of trust and mutual respect that is more apparent than any other place I have worked. There is an expectation that something positive will happen. The total buy-in by...our industry partners has been the key to Arizona's success."

Partnering Applications Other Than Single-Project

At first I thought this book would focus only on single-project partnering. However, I skimmed through the last three years' worth of issues of the newsletter, *Contractor's Business Management Report* (which I edit), and singled out what I thought were the most significant articles. A common thread emerged: Each involved partnering, in most cases firm–client, or employer–employee. In some cases the goal was safety, in others, superior contractor performance, or equal opportunity. Thus the focus of the book became much broader than just single-project partnering.

Multiproject Strategic Partnering

Contractor management consultant Anthony Costonis, of Corporate Development Services (Lynnfield, Massachusetts), has subcontractor-clients doing construction projects with three of America's premier industrial firms: DuPont, Ford, and Intel. Thus Costonis has an extraordinary vantage point from which to write about multiproject strategic partnering. In Chap. 8 he reports that DuPont, Ford, and Intel are all taking much the same approach with their contractors. This means:

1. They are screening contractors, typically narrowing their number by 90 percent or so.

2. Thereafter, rather than bidding out each new project, they use the same suppliers job after job. Owner and contractor form a continuing team. At each job's end they study what went well, what did not, and why. They take those insights to the next job, repeat the things that went well, try something new where there were problems.

3. Commonly, such vendors tighten the screws on contractors: "Within a year, your work must come in with unit costs 10 percent lower" (or whatever). But not infrequently they have learned to work so much more efficiently, despite the financial constraints, that contractors are nonetheless making more profit.

So far this approach has been used mainly by very large industrial owners. In 1994, however, the Construction Industry Institute reported that strategic partnering saves owners an average of 15 percent, so we can expect more of it with clients and jobs of all sizes. To enjoy these savings, however, you must adopt a quality program (such as TQM) or raise your quality quotient in other ways.

Lawyers, Claims, and Litigation

The attorney's role

During the 1970s and 1980s, the best-read column in *Civil Engineering* magazine, to the dismay of its editors, was "Court Decisions." During that same period the number of attorneys specializing in construction law skyrocketed. Why? The 1992 president of the AGC, Marvin Black, head of Marvin M. Black Co. (Atlanta), says it was the "era of abundance" of the past several decades. Both contractors and owners had it relatively easy, and "[w]e got fat and lazy." When trouble hit a construction project, the tendency was to turn to the attorneys.

When Black visited the AGC's local chapters, he would ask his audiences: "How many of you paid your lawyers more than you made in profits last year?" About 10 percent always raised their hands. And were he to ask how many were involved in litigation, he estimated, about 80 percent would raise their hands.

Black predicts that this will change. "Already, owners can call up on their computers, public-domain data on the safety records of contractors. I think the same thing will happen (if it hasn't already) regarding public records showing how litigation-prone contractors are." Owners will choose contractors, in part, based on how litigation-averse they are.

As a result, writes attorney Fred Lyon, of the law firm Lyon & McManus (Washington, D.C.), the role of construction lawyers will be transformed (see Chap. 9). Lyon challenges both lawyers and their construction clients.

- Lawyers: Ensure that contract documents allocate risk equitably. Attend the prejob partnering workshop to listen to the parties' mutual objectives. Tell the parties you will consider that they have failed if disputes cannot be avoided/resolved, and litigation results. And if the client resists partnering, Lyon says, tell them to get a new lawyer.
- Contractors, owners, and design firms: Lyon says that if lawyers are not included in the process, they will work to destroy it. For

instance, if lawyers try to "serve" you by shifting risks you should be handling to other parties in the contract, those other parties' lawyers are likely to attack you on that basis.

Alternative dispute resolution

Two other construction lawyers, Robert MacPherson of the law firm Postner and Rubin (New York City) and Richard Steen of the contractor consultant Hill International (Willingboro, New Jersey), tell us in Chap. 10 that partnering is not really revolutionary: It is merely the ultimate form of alternative dispute resolution, which began decades ago with arbitration and evolved into mediation and several other time- and cost-cutting alternatives to litigation. Partnering is just the next step.

Steen describes one of the newest variations in alternative dispute resolution, the "Project Neutral." Having such a person on the team makes an alternative dispute resolution tool available before a dispute starts—so it is less likely to happen at all, or if it does, to have serious consequences.

The Project Neutral is a one-person dispute review board, and because only one person is involved, it may be less costly than other means. A well-chosen Project Neutral may have two advantages over many partnering facilitators: he or she is on call throughout the job and not solely as a retreat facilitator, and he or she will have construction technical, management, and/or construction finance and management competence, which some partnering facilitators lack.

Partnering between Management and Employees

Why Colorado steel-erection contractors are safety leaders

Construction management consultant Steve Miller of Miller Safety (Thornton, Colorado) has helped Colorado steel-erection contractors transform their business. The industry needed radical surgery: It was so unsafe that workers' compensation insurance cost more than the steel erectors' base pay. Among the changes that have been made in the last five years are the following:

- The accident rate has been cut by almost half.
- A steel-erector "tie-off" system now permits workers, even on the top steel of a building going up, to work safely.
- The approach that OSHA Region 8 (Denver) and the Colorado steel erectors used—called negotiated rule making—is being tried in

rewriting an OSHA standard on the national level. OSHA alone is no longer drafting a standard and then asking for comment. Instead, OSHA, steel erectors, steel workers, and others are doing it jointly. In Chap. 11, Miller tells how it all began.

Why BE&K Construction is a leader in honoring employees

Attorney Nick Bouler, personnel chief of the large industrial contractor BE&K (Birmingham, Alabama), recalls in Chap. 12 how his company, relatively young among major contractors at 20-plus years, came to emphasize the "honor our employees" idea: Twenty years ago, when most construction labor was union, BE&K was a pioneer in being nonunion. If cofounder Ted Kennedy was to recruit and keep good workers, he would have to offer them as much or more than the unions did. Bouler says:

> Kennedy has lasting memories of the days when the very continuation of the company depended on getting craft workers to travel to a project to do quality work for a (then) unknown company. In those early years, recruiting those hands and their field supervisors was his job, and he's never forgotten the men and women who accepted the challenge to help him make this upstart company into an industry leader.

The marching orders were, and are: "Our assets walk out the gate every night. If we haven't provided a job they want to come back to tomorrow, we're out of business."

How Parsons Brinckerhoff built one of the largest international practices of any U.S. engineering firm

Chairman Henry Michel may have more foreign design-firm experience than any U.S. counterpart, and it shows: The firm has 500 employees overseas, 360 in Hong Kong alone, and 30 ongoing jobs in China. In Chap. 13, Michel relates some of the steps in building to this leadership.

Perhaps most telling is his recounting of how the Hong Kong office grew so big:

> With access to upper management jobs, Chinese were able to contribute creatively to the technical and marketing sides of the business, to receive credit and rewards for their contributions, to take pride in the work done, and to become shareholders....
>
> By entrusting responsibility to Chinese project managers the firm gained a cadre of effective new leadership with invaluable connections to local clients.

Most important, it gained a core staff ready to export their services to the other Pacific rim areas. That group also became the basic cadre to open up new companies in Singapore, Malaysia, and Thailand. Those companies are totally locally staffed and managed by transplanted Asians whom we had trained at their home base. Singapore now has 100 employees, for example.

To close the loop, Asian-trained PB engineers/stockholders are now helping staff work at JFK International airport in New York, the Cairo Metro in Egypt, and the Jubilee Line of the London Underground.

Partnering with Customers and Employees via Total Quality Management

Bruce Woolpert is president of Granite Rock Co. (Watsonville, California). His is the first construction industry company to win the Malcolm Baldrige National Quality Award. In Chap. 14, Woolpert tells how he did it. The secret was partnering with Granite Rock's employees and customers—that is, serving them more aggressively and imaginatively than competitors did.

Ron Deffenbaugh is vice president for quality for contractor McDevitt Street Bovis (Charlotte, North Carolina). There, as he reports in Chap. 15., TQM is pursued aggressively, and it is paying off: Legal fees are down about 75 percent in four years. The firm remained profitable during 1990–1994 despite changing hands, and despite the recession.

One TQM key is to measure your processes. In the past this meant primarily measuring two things—costs and schedule. Under TQM, the firm today is measuring much more, including the monitoring of conformance to schedule of:

- Requests for information process—number submitted with solutions, number submitted versus returned on time.
- Submittal process—percentage submitted on time, and percentage returned on time.
- Payment process—days overdue by owner, days overdue by general contractor.
- Change order process—percentage on time pricing turnaround, accuracy of pricing.

Partnering between General Contractors and Subcontractors

Lorry Bannes is president of Bannes Consulting Group (St. Louis). In Chap. 16 he tells how an inner-city St. Louis medical clinic, though built largely by minority subcontractors and almost entirely by minority labor, was brought in both on time and on budget.

The secret? Like most subcontractors, St. Louis' minority subs knew their trades but were relative greenhorns in business management. Bannes became their teacher, cheerleader, and mentor. Before and during the project he held formal and informal classes in estimating, scheduling, project manning, and other areas. He has built 1000 projects over a long career, and counts this one among the most satisfying.

Gerry Graff is president of Graff Flooring (Albuquerque, New Mexico) and 1994–1995 president of the American Subcontractors Association (ASA). In Chap. 17 he recalls that previous efforts at writing a model general contractor–subcontractor contract failed. In 1994, however, representatives of the two groups succeeded. No surprise here: The general contractors and subs partnered. Their success stands in sharp contrast to former attempts, when the AGC prepared a proposed model subcontract and asked the subs to approve it. Not surprisingly, they did not.

A Look at Tomorrow

Finally, this book offers two visions of the future: Both of these authors believe that project partnering's applications will stretch beyond single- and multiproject partnering. For example, management consultant, lecturer, and architect Kyle Davy (Berkeley, California) says in Chap. 18 that partnering has taught three lessons:

1. *The wisdom of teams.* On every project, the A/E/C community says they have a "project team." But in most cases it is really a pseudo-team—its "members" have little interest in shaping a common purpose or set of performance goals. Trust is virtually nonexistent, and individuals act according to their own agendas.

Partnering builds real teams. Its members all commit to a common purpose, performance goals, and approach, for which they hold themselves mutually accountable. Team building teaches them to trust each other, to communicate openly, to solve problems quickly.

2. *A commitment to learning.* Partnering encourages organizations to learn. See especially the two chapters (14 and 15) on TQM in construction companies. Also, see Nick Bouler's chapter on BE&K (Chap. 12). Firms grow by "growing" their people, and by reforming what their people do on the job.

3. *Breaking the rules.* After a decade of TQM and incremental improvements, some contractors are realizing that improving existing processes is no longer enough. For example, in refitting a large existing office complex, R.J. Reynolds Tobacco (Winston-Salem, North Carolina) did away with contractor bidding and virtually eliminated

contracts. Instead of using the sophisticated management control systems usually needed to manage a large project, they split each job into small work packages of 7000 to 12,000 ft^2 each. Work was awarded on a time and materials basis. Multiproject partnering allowed owner–subcontractor teams to move up learning curves. As a result, construction cost and time were cut as much as 50 percent.

The author of Chap. 19, Jerry Pitzrick, is a partnering consultant with Mortenson Construction (Minneapolis) and has facilitated over 70 preproject partnering retreats. Since some have been for Mortenson jobs and others for other clients, Pitzrick has an unusually broad view. That is probably the reason for his breathtakingly bold visions of possible futures. Pitzrick draws on what he has heard from owners, architect/engineers, and contractors, and in researching this chapter Pitzrick also picked the brains of building owners, construction labor specialists, and others. Pitzrick believes that partnering could be an opening wedge to transforming the way construction is done.

Conclusion

"Would you partner again?" contractor management consultant FMI (Denver) asked 114 contractors who had used the process. "Yes," answered 92 percent, according to FMI's Lou Bainbridge (Chap. 7). The four who answered "No" cited these reasons:

1. Bad weather had been a factor.
2. Trust/respect was poor.
3. There were not enough "take-charge" people in critical decision-making roles.
4. Participants had been "forced" to partner.

Except for weather (which is irrelevant), all other negative responses had to do with leadership, or lack of it. That factor is just as critical to partnering success as is mastering the tools of partnering. I believe that this book's 21 contributors personify leadership. They certainly built this book.

Chapter 2

How Public-Sector (Fixed-Price) Partnering Got Started

Larry Bonine

This chapter discusses the genesis of public-sector or fixed-price partnering. In a way, this is the story of why and how partnering got its industry enthusiasm and momentum.

Larry Bonine, director of the Arizona Department of Transportation, recalls the advent of single-project partnering in 1988 at the Oliver Lock and Dam project in Alabama. As the partnering workshop kicked off, each participant was asked to fill out a nine-question questionnaire. The inquiries get at the guts of partnering.

The term *partnering* originated with the private sector and has actually been around for years. Private-sector partnering is in fact more of a strategic alliance formed by two or more companies for mutual benefit and does not have much to do with claims, disputes, or lawsuits (see Chap. 8). What we have in government now is a partnering that has been designed to send the problem back to where it belongs, back to people—the owner of the constructed facility and the supplier who designs and builds it. I guess we have stolen the word, but partnering is now our word, and here is where it came from.

"I was never ruined but twice: once when I lost a lawsuit, the other time when I won one."

VOLTAIRE

The construction industry was, and is, experiencing a growth in claims and related litigation that is without precedent in our history.

All too often, an adversarial relationship between the owner and the contractor is not only expected but is considered natural and even acceptable. Today, we have stopped talking and listening to each other face to face and, in too many cases, are relying on letters and formal communication to interact.

Companies and consultants are emerging from every nook and cranny to either help create a claim situation or to help an owner mitigate claims. With so many people earning a living based on claims, it is no wonder that adversarial relationships emerge between owners and contractors. It is also no wonder that partnering is meeting such a willing audience. Owners and contractors are, quite frankly, fed up with having to use the legal process to solve their problems. They are finding the partnering process to be an answer, an alternative to traditional adversarial conflict.

Partnering is all about people. People have gotten us to where we are in the construction industry today, and people will be the answer to an improved way of doing things in the future. People are the most important and most complicated part of the process. Partnering works because it is one of the few quality programs that is all about people. If people understand each other, develop a mutual respect and try to succeed together in a partnered project, they will.

Construction of major facilities such as highways, tunnels, skyscrapers, and dams is the ultimate manufacturing process. Each facility is constructed only once. Continuous improvement can happen along the way, but you can never go back except to correct mistakes. It is a one-time process—once built, it is finished for life, unless expensive, disruptive rebuild is undertaken. The trick is to build it right the first time.

When a crane or other major piece of equipment needs replacing, it is usually apparent, and the system makes it happen immediately so that production and scheduling milestones can be met. The same is not true in how we deal with the most important part—people. Our managers, especially our loyal managers, can mess things up royally, and we will hang in with them and hope the problems will sort themselves out. We will support them at all costs. Admirable but not awfully smart.

Let's go back and see how we selected this manager in the first place. Did we select him because he demonstrated skills in working with people? Did we select him because he showed a desire and ability to get things done through people? Did we consider for a second what his potential style of management might be? Or did we make our selection based on engineering skill, hard work, and seniority? In the Corps of Engineers, the last is all too often the case. In the Corps, we could not even give a young bright engineer leadership training because the human resource section said this might look like preselection. Those managers doing the selecting could only look for candidates who were good technically and who had perhaps led committees, been in Toastmasters, and demonstrated an interest in working with people in other ways.

So now in our industry we have some managers who may need serious leadership training and a lot of self-confidence if they are going to answer the call of enlightened leadership and empower their people to make decisions and even make mistakes. It takes a confident leader to turn it all over to a team and allow subordinates the freedom to fail. This is the key reason that partnering has been successful: It emphasizes people, teaches communication skills, and empowers a person in problem solving.

Dan Burns Introduces Partnering to the Mobile District

Some time early in 1988, the chief of construction for the Mobile District came to see me about something he was considering for our district. Dan Burns was the least likely looking chief of construction you ever met. He was tall, slim, articulate, and not from the South. I had always pictured construction types a little bit differently.

After the usual pleasantries about golf, Dan got deadly serious and wanted my complete attention. In this one-on-one setting he laid out an exciting vision of what he called partnering in a fixed-price arena. He felt that what was working in the private sector as an alliance could work on the public side. By the time his presentation was through, he had me hooked. It had to work. Why hadn't we been doing it all along? Right then, I became a cheerleader and a passionate advocate for partnering.

Within an Army Corps of Engineers District the size of Mobile, there is always a lot going on. We were a lot like a private engineering firm in that workload was important to us. As district engineer (the boss), it was my job to get new business and foster customer relations. Our most sensitive customer was the U.S. Air Force. By law, we were responsible for design and construction of their facilities. They were a captive customer, and unfortunately we behaved accordingly.

The Arnold Engineering Development Center (AEDC) at Tullahoma, Tennessee, was the home of an element of the Air Force Systems Command. We had participated in numerous major projects at Arnold. Unfortunately, the last one, a gigantic wind tunnel, was not a proud moment for the Air Force or the Corps. Everything that could go wrong did go wrong. We experienced design problems/disputes and construction problems/disputes. The Corps was unhappy and our customer, the Air Force, was unhappy. In addition, the Air Force had an operating contractor at Arnold that was capable of building virtually anything, and we believed the Air Force could have made a strong case for building future projects in-house—i.e., not using the Corps for their design and construction services. Treating the Air Force as a captive customer would no longer work. We had to do something proactive to keep AEDC.

The Large Rocket Test Facility (LRTF/J6) was under design when I arrived at Mobile, but there was still discussion about who would be construction manager. We wanted very much to impress Systems Command as to our ability to make the LRTF the best project they had seen. We adopted partnering as a technique to better serve the Air Force and to maximize our success potential for J6. Construction of J6 was still a ways off, however, and we wanted to start partnering immediately—to try it out. The logical choice was our next major project, Oliver Lock and Dam.

Oliver Lock and Dam

The first federal fixed-price, low-bid contract in which partnering was used as a vehicle to facilitate communication between the government and a contractor was the second phase of construction at Oliver Lock and Dam in late 1988. The partnership would be rather simple, since the Corps was the owner, designer, construction manager, and operator. In those days, I would never have used the term "simple" to describe a partnership, but today we have partnerships with as many as seven different key participants.

The low bidder, by about $7 million, was FRU CON. FRU CON is a German-owned and -managed company. The director for U.S. operations and the leadership of the construction project were all Germans and highly qualified.

Being in Mobile, Alabama, and having a continuing relationship with the University of South Alabama, we were introduced to a couple of professors who had impressed us with their ability to facilitate people into creative ideas. We asked for their help and ideas in putting on a workshop to create a model for government partnering. We decided on a three-and-a-half-day workshop to accomplish several things:

- Team building
- Problem solving
- Developing a model for escalating problems quickly

What we got was much more.

The session was held at the conference facilities at the university. The participants were asked to stay in the facility overnight and to socialize in the evenings at dinner.

For this first-ever government partnering effort, only the absolutely key actors from both sides were present. I know now that many more partners/stakeholders should have been brought on board. As district engineer, I kicked off the workshop with what I thought was a dynamic and inspiring speech on teams and the advantages of participatory management. I was followed by Manfred Lupp, the president of FRU CON's U.S. operations. In his accented but understandable and articulate English, he promised his full support of what we were trying to do.

Following this brief but significant kickoff, the partnership went to work. One of the first things the facilitator did was ask each participant to complete a nine-question questionnaire. Names were not requested; however, respondents were asked to note whether they were contractor or Corps. The questionnaire looked something like the one shown in Fig. 2.1. The questionnaires were collected, put away, and not mentioned again for three days.

Now the facilitators started to earn their money. While a truly good facilitator should be invisible, a certain amount of leadership is required to get people to start interacting. Don Mosley and Carl Moore are brilliant at doing this. The emerging partnership begins with some hands-on experiences that demonstrate how powerful team play can be. Properly introduced and administered, these exercises set a tone that keeps everyone's attention for the rest of the workshop.

With the participants loosened up, the serious work began. Each side separated for what was to be the last time. They were asked to develop their goals for the upcoming construction project. Also, they were asked to list the other side's goals. This was pretty serious stuff. Many of the participants on both sides had never been asked to articulate their goals for a project. This was their first taste of empowerment, and from the feedback they liked it.

The next phase of this first workshop zeroed in on people. Using the Myers Briggs Test Indicator (MBTI), the shocking realization that we are very different was brought home. Teams, as well as managers, must realize that we are all motivated differently. If you treat everyone the same, you will get mediocre performance at best. I know of no better tool than MBTI to demonstrate just how different we are.

During the MBTI exercise, putting people first begins to have meaning. The participants of every partnering session in which I have participated seem to value their experience with MBTI as much as anything else we do.

I am with:

Owner____ On site____ HQ____

Contractor____ On site____ HQ____

PRE-WORKSHOP SURVEY

Based on my experience with or knowledge of public construction projects like the upcoming OLIVER LOCK & DAM Project, I think it's typical that:

1. Communication between contractor and owner personnel is:

1   2   3   4   5

Difficult, with much misunderstanding   Open, honest, flowing

2. Concerns and problems are acknowledged:

1   2   3   4   5

Only when they can't be ignored   At first

3. Concerns and problems are:

1   2   3   4   5

Swept under the rug   Dealt with quickly and directly

4. Cooperation between owner and contractor personnel is:

1   2   3   4   5

Non-existent   Characteristic of all phases of work

5. When issues are raised, *our* response is:

1   2   3   4   5

Extremely slow   Prompt & responsive

6. When issues are raised, *the other guy's* response is:

1   2   3   4   5

Extremely slow   Prompt & responsive

7. When issues are raised, people:

1   2   3   4   5

Say one thing, but do another   Do what they say they will do

8. When projects are completed, a sense of teamwork between owner and contractor staff is:

1   2   3   4   5

Non-existent   Strong

9. I define a successful project as:

Figure 2.1 Pre-Workshop Survey used at the Corps of Engineers' first partnering workshop, at the Oliver Lock and Dam project in Alabama in 1988.

After the "people" work, the partnership went to work on a dispute model. We did not know it then, but a significant key to the success of partnering is to agree before you start as to how to resolve any disputes. For this first effort, we probably could have been more successful in developing the problem-solving model.

Another important product was to decide in advance to have follow-up workshops. The partnership agreed to meet and review performance periodically, a sort of health check to see how things were going.

Finally, the partnership developed a partnering agreement and a joint logo. The agreement was signed by all members, and the logo was worn on the hard hats on site by both the Corps and FRU CON. Not a real big gesture, but if you know anything about the Corps of Engineers, you know they are proud of their castle and using another logo was considered a major step toward partnering.

This first partnership had a lot of problems. While it started very well and accomplished all its immediate goals and objectives, we could have done a better job. I am not sure we recognized how much work nurturing a partnering relationship takes.

That the concept of public-sector fixed-price partnering came from the Corps of Engineers is almost unbelievable. The Corps construction program (and this goes double when you talk about the Mobile District) was so control oriented that I still cannot believe the Corps started this bold new initiative. I had contractors tell me that they did not bid our work, or if they did, they added a significant "Mobile District fee" to any work they bid on. With that reputation and the first partnership beginning with what the Corps considers its primary specialty, a lock and dam, I was elated that it worked as well as it did.

The key to success on this project was having the new lock constructed and opened with minimal impact to the Black Warrior-Tombigbee Waterway. The project was in the district of the chairman of the Energy and Science Subcommittee of the House Appropriations Committee, Tom Bevil. To keep barge traffic flowing would require blowing up the existing lock and dam and transitioning to the new lock within three weeks. Without going into details, let me just say this was a significant task that had the interest of the entire region. Partnering was a key to the success as we met this milestone.

Dampened Success

With the changing of the guard in the Mobile District, in terms of both military and civilian leadership, and with a lack of enthusiasm, command emphasis, and an identifiable champion, I understand that the partnership did not continue to enjoy the success of its first two years. That is unfortunate, as the Oliver Lock and Dam played an important role in shaking up the construction industry. First, it was the first attempt to put the people part first. Second, it was the catalyst that got the U.S. Air Force interested in partnering. Third, and perhaps most important, the Oliver Lock and Dam got Chuck Cowan turned on to partnering. Both the Air Force and Chuck Cowan are major players in the partnering movement, and enough cannot be said about the importance of their participation.

Chuck Cowan

Charles Cowan was district engineer of the U.S. Army Engineer District, Portland, when he was introduced to partnering (see Chap. 3). The district was charged with building the Bonneville Lock and Dam. In gearing up to meet the challenge of major lock and dam construction, Chuck heard about what was happening at the Oliver Lock and Dam. He took the time to bring members of his staff to observe first-hand how partnering was working. Both the contractor and the construction management staff reported positive results, enough so Colonel Cowan decided to use the concept as another tool to help manage his complex project in Oregon.

The Northwest was ready for the concept. The contractor not only joined the district in partnering but became an enthusiastic participant. The project was so successful that it became a model and deserves credit for demonstrating the potential of owner and contractor working for mutual goals. The industry watched and accepted partnering, and its new champion spread the philosophy.

Chuck Cowan was so successful that the governor of Arizona asked him to join him in bringing partnering to state highway construction. The opportunity to make a difference in the industry was just too good to pass up. Chuck accepted and joined the Arizona Department of Transportation (ADOT). Under the leadership of Chuck Cowan and others in the department, such as then Chief of Construction Tom Warne, Chief of Operations August Hardt, and State Engineer Gary Robinson, the department began to include partnering in all state contracts. Other states are following Arizona's lead; at last count fully 85 percent of state DOTs use partnering to some degree. While I may have been involved in the "first" public partnering effort, I recognize Chuck Cowan as the person who really got the industry to accept it.

The U.S. Air Force

A few months after the Oliver Lock and Dam was under way and some encouraging results were coming in, we in the Mobile District had one of our periodic meetings with our military customers, the Air Force Base Civil Engineers and the Army Directors of Engineering and Housing. These were general briefings and feedback sessions, where we tried to open ourselves to our customers and to improve the services we were providing. During the closing part of the two-day session, I explained what we were doing with partnering at Oliver. The audience listened, and during the Q&A session another milestone on the road to partnering acceptance occurred.

Joe Hartung, then a contractor for the Air Force and in charge of the Cape Canaveral construction program, asked: "Colonel, why don't you do this on our jobs at the Cape? If you do it for yourself, why don't you offer it to us?" There was silence as I considered the question, but it was not a long one. We agreed then and there to partner the complicated Tactical Operations Control Center and the huge Solid Motor Assembly Building. These jobs were imminent and were both destined to be immensely successful. They allowed for multiple stakeholders to become involved and to accept that we may have happened upon a better way of approaching design and construction.

Joe Hartung was and is a real crusader, who embraced partnering on first hearing about it at that key closing briefing in Mobile in 1989. He is now a skilled and innovative partnering facilitator and coach, practicing in the Cape Canaveral area, and must be noted as playing a key role in our success and in the success of partnering today.

Why Partnering Works

Something must be said in any book on partnering on why it works. Here are my thoughts.

Partnering puts the people part first. There are simply no technical problems that people cannot solve. It is the people part, however, that also causes most of the confusion. We have the potential to do everything wrong: Egos get in the way, we don't listen, we posture, we misunderstand, we know what we know and just don't take the time to find out what the other guy might know, we don't try hard enough to put ourselves in the other person's shoes, and we forget the golden rule of reciprocity.

Partnering requires that we exhibit both courage and consideration. We must have the courage to present our side of the story and speak up for it, but we must be considerate of others and their interests as well. We have to know and care what happens if someone loses out and gets nothing in the resolution process. In our industry, we must have two winners. None of us can stay in business and continue otherwise. Partnering is about understanding where the other person is coming from. Listening, caring, and questioning, genuinely trying to grasp the comprehensive situation are things we *think* we practice, but alas, my experience has been that we don't work very hard at it.

Few of us learn much about listening in school. We are taught reading, writing, and how to speak properly, but the most important learning skill, listening, just slips right by as a taught skill. Only when there is a crisis do we sometimes take the time to understand—and by crisis time it is often too late. Although some companies are beginning to sponsor training in listening skills, we are just now starting (often through partnering) to grasp the scope of the potential benefits.

The bottom line is that partnering works; there is overwhelming proof and testimony to convince all but the most closed-minded and cynical people that it is worthwhile. At the Arizona Department of Transportation, in mid-1994 we held our annual Materials and Construction Conference. Attendees included construction and inspection people, contractors, subcontractors, and representatives from the Federal Highway Administration—in short, representatives from across the construction community.

Although the conference lasted two days, the highlight this year was the first annual awards banquet. At this grand affair, each of the 10 districts was invited to recognize their best projects. While this showcase will never replace Hollywood's Academy Awards ceremony, there was a certain amount of pride and excitement as each district engineer in turn named a project to be recognized. The resident engineer, the contractor's project manager, and the designer came forward together before an audience of 300 to receive their plaque and have their project recognized.

Although this was neither a polished nor rehearsed event, the words of the recipients were significant. Every one of them talked about partnering. They used the word *trust* a lot, and that term is not commonly used in the construction industry. From the stories I have heard about the ADOT, the word had not previously been used very much there. But trust, trustworthy, respect, and teamwork were used repeatedly by all of those who were recognized. They said things such as, "it was fun to go to work on our project," and one contractor said that "if there was ever any doubt that *partnering works,* you should have visited our job site." All 300 people listened and watched attentively, but I could tell they were not surprised. Partnering is just the way we do things in Arizona now. As I watched, I thought that this event alone proves that we have something going here that is powerful and is having a major impact in the world on engineering and construction.

In Arizona, our industry is starting to be a good place to work again. We have had over two years of partnering experience without having to use "legal" remedies. By the fall of 1994 we had completed 96 projects and had 88 in progress. We partnered every one of them, no matter how big or how small. We have always included the designer, since almost every decision comes from an interpretation of the plans or specifications. We have empowered our resident engineers and the contractors' project managers so they can make final decisions. Therefore, the problems are resolved on the job site, at the level closest to the issue.

Partnering Can Change Whole Organizations

Once an organization embraces the concept of people first, empowers them, and goes with a partnering initiative, that organization will never be the same again. The organization starts a growth and development process that begins to involve everyone in everything. As people attend partnering workshops and experience the synergy and magic of the human factor, they come back to their section or branch in the home office and ask, if it will work with an outside contractor why won't it work internally? Why won't it work here? In addition, they begin to use some of the skills of team building and problem solving they learned in the workshop. They find that these techniques work and want to share these powerful tools with others. A multiplier starts to unfold, and soon the whole organization starts changing toward empowerment of people and participatory management.

I have had personal experience with this phenomenon in Mobile, at Boston's Central Artery/Tunnel, and now at the ADOT.

The Future

Where do we go from here? I can tell you where the Arizona Department of Transportation is going. We will continue partnering with our suppliers, we will continue to address problems as they come up, and we will work with suppliers and encourage them to adopt our philosophy and quality initiative. The future of partnering (which in my opinion is the greatest paradigm shift that has occurred during my lifetime in the construction industry) is toward the total quality initiative that our country's industries have had to embrace to survive (see Chap. 15). Partnering follows (or leads) the total quality management principles so well that engineering and design companies find quality and partnering go hand in glove.

Chapter 3

A Contractor's Partnership Baptism

Michael Murphy

Breakthrough solutions are most likely to result from bold visions or huge problems. The latter was the spark that gave birth to single-project partnering in public works, according to Michael Murphy of Donald B. Murphy Contractor (Federal Way, Washington).

Charles A. Cowan, then a colonel in the U.S. Army, describes his first morning as commander of the Corps of Engineers' Portland (Oregon) District. He arrived at the office early, wanting to make a good and lasting impression. While he was making the rounds introducing himself, he felt a tap on his shoulder. As he turned to shake hands, he was presented with a Summons and Complaint by a process server.

David Johnson, a member of the Corps' in-house legal staff, tried to reassure Col. Cowan: "There's nothing to worry about. This happens all the time. You can plan to spend a large part of your day working on a variety of pending litigation." [Editor's note: A Corps attorney in Washington, D.C., told me in 1991, "I have $1 billion in construction claims on my desk."]

Chuck Cowan was not satisfied with "getting used to it." He decided there must be a better system for administering public works contracts and avoiding disputes. We now recognize the system he developed as partnering.

This is the story of how partnering was used to solve many critical problems in one relatively small part of a massive construction project. The small but critical piece of the project was the tiebacks that held Union Pacific Railroad tracks in place on an embankment near the Bonneville Lock, and that also had to hold the Columbia River out of the lock until the appropriate time. While the subcontract, at $4 million, was not huge, its meaning to Donald B. Murphy Contractor, Inc. (DBM) was and is profound. It has forever changed the way DBM looks at the construction business. Moreover, it is the story of a group of people taking control of one miserable job and making it fun.

The Bonneville Navigation Lock Project

Built in 1938, the Bonneville Lock, 500 ft long and 76 ft wide, is the smallest of eight locks on the Columbia River. Because of the lock's relatively small size, it was necessary to break a multibarge tow into smaller units, pass each unit through the lock one at a time, and reassemble the tow at the other end of the lock. In short, the lock was a bottleneck.

To increase the capacity of the lock, it was decided to build a new one, substantially longer and slightly wider at 675 ft by 86 ft. The additional length would permit an entire, multiple-barge tow to pass through the lock. The recently completed Bonneville Navigation Lock project, costing $331 million, makes this possible, and unplugs the bottleneck.

Partnering Concepts

What is partnering, who are the partners, and what is a dispute?

- Partnering is a simple, common-sense approach to dispute avoidance.
- Partnering involves proactive action to address project problems.
- A dispute is nothing more than a problem which was never solved.
- A stakeholder is any person, group, entity, or organization which has a vested interest in the project.
- Partners are or should be the stakeholders in the project.

Partnering will never eliminate problems. It is important to recognize that there is no sin or disgrace in having problems on a project—they have existed and always will.

Partnering creates a mechanism for solving project problems. If you understand and apply five simple partnering concepts—leadership, empathy, equity, trust, and honoring the spirit of the agreement—to the construction and administration of a project, your chances for success will be increased dramatically. The practice of these concepts is not easy. Partnering takes work!

Leadership

Partnering is proactive. In partnering, people take control of their destiny by action rather than acquiescing to being controlled by events and others.

Partnering starts with leadership. Leaders control their own destiny as well as that of those who are more comfortable in the role of following. Leaders do not like to sit back and constantly react to their environment; they want to shape and manage it. By the fact that you are reading this book, one can assume that you may not be totally satisfied with the current state of your professional environment. One might also conclude that you want to do something about it. This act, by its very nature, shows that you are taking control of the situation, as all leaders do.

In partnering, the single most important thing an effective leader does is empower those under him. An effective leader empowers the hierarchy from top to bottom to make timely decisions. The leader requires no one to make a decision that he or she is not comfortable making, but encourages decision making at the lowest possible level. A real leader gives both the authority and responsibility to make decisions to workers at all levels of the project.

Empowerment brings the whole organization on line to achieve the common goals. The key is to gain the participation of the entire hierarchy, for each individual stakeholder has always had the ability to short-circuit agreements. It is imperative for leaders to recognize this potential roadblock to success. When leaders empower their subordinates, they support those below them who make decisions. True leaders support their coworkers and never reverse the roles. The Bonneville project had many examples of leadership.

Charles A. Cowan is often called the "father of public works partnering." Then a colonel in the U.S. Army, Cowan describes his first morning as commander of the Portland Oregon District Office of the Army Corps of Engineers in the following terms. He arrived at the office early, wanting to make a good and lasting impression on his new command. As a commander in a district office of the Corps, he would be primarily commanding civilian civil service employees of the federal government. His uniform, though impressive, would mean less to the civilians than to a group of subordinate officers and noncoms. Respect would be slightly less of a given. Hence, he was applying all of his charm and charisma as he introduced himself.

As he was making the rounds, he felt a tap on his shoulder. As he turned to shake hands with what surely must be another member of the staff, he was presented with a Summons and Complaint by a process server. Once Cowan had waded through a variety of Latin phrases, he was profoundly confused. David Johnson, a member of the in-house legal staff for the Portland District, tried to reassure Col. Cowan and said, "There's nothing to worry about. This happens all the time."

Mr. Johnson went on to translate the document: "Boss, this says by virtue of your position in the Army Corps of Engineers, you have behaved in an arbitrary and capricious way. It goes on to say that the aforementioned behavior has virtually bankrupted the contractor. It says you have breached the contract in a variety of ways. To sum it all up, it says the government owes him a large sum of money and additional contract time. You can plan to spend a large part of your day working on a variety of pending litigation."

Chuck Cowan was not satisfied with "getting used to it." He decided to do something about it, and he put Mr. Johnson in charge of practicing "preventive law." Cowan decided that there must be a better system for administering public works contracts and avoiding disputes. He empowered the rest of the staff to create a model system to prevent the destructive litigation that had become so pervasive in construction. We now recognize that system as partnering.

Leadership on the Bonneville project did not stop with Col. Cowan. As it happened, a portion of the lock construction needed to be completed within the Columbia River. This work had to be done during a very restricted time period between the several salmon runs which pass by the dam each year. For this work to be completed, the excavation in the upstream area of the lock had to be finished. And for that to be done, a series of very difficult tiebacks had to be installed through two concrete "slurry walls." These underground walls, installed under a previous contract, were designed to keep the Union Pacific Railroad on the tracks and the Columbia River out of the excavation, even after the earth was excavated on one side on the walls. All of this added up to a very difficult "critical path": A *one-day delay* could translate into a *one-year delay* waiting for the next construction window, between fish passages, to open.

Enter now the skeptical, and by some measure recalcitrant, tieback subcontractor Donald B. Murphy Contractor, Inc. A variety of problems related to the installation of the tiebacks occurred immediately. Most of them had negative schedule impacts. Some of the problems were owned by the subcontractor, some by the general contractor, and some by the Corps of Engineers. The situation quickly deteriorated into the creation of the paper trails needed to prove that each of the stakeholders was more right than the others *when*, not *if*, litigation started. Rome was burning.

Nero, however, was not picking on his fiddle. Nero in this case was Richard Geary, president of the general contractor on Murphy's job, the Kiewit Pacific Company. Dick Geary was and is a true leader. He took control of the destiny of the project. He called a meeting with Donald B. Murphy Contractor, DBM's counsel, the Kiewit/Al Johnson (KAJ) senior project staff, and himself. It was the quintessential "come to Jesus" meeting, to use the construction vernacular.

By way of background, DBM was not under contract when the original facilitated partnering session took place, so they did not participate in the preproject partnering retreat. When DBM arrived on site, they were given a copy of the partnering charter and informed that they would be part of an agreement they had no part in creating. The recalcitrant and very wary subcontractor read the charter and immediately commented, "This is a novel way to dupe and defraud the contractor."

To continue, Dick Geary, during the meeting, made a convincing appeal in the form of several not-so-veiled threats, that the partnering charter was for real and the words were not empty. He informed DBM that the subcontractor's desire to "get paid for his work," a euphemism for being claims oriented, would make the project very miserable for all. He took charge.

There was no question that Geary meant what he said. He continued the discussion by informing those present that the project staff had both the carrot and the stick necessary to bring DBM into the fold. Perhaps the most important thing he accomplished was to empower the reluctant subcontractor to control its destiny. It was clear that there were two paths that could be taken—partner or litigate. It was also clear that only the first made any sense. Here began the metamorphosis of DBM, from standing for *Don't Blame Me* to understanding that partnering offers a welcome prospect for long-term business survival in a litigious world. *Metamorphosis* is the operative word: The transformation was by no means complete, but at least the process had begun.

The above anecdote is provided to show not only leadership but the importance of including all the stakeholders in the partnering process. Without Richard Geary's intervention, regardless of who owned which problems, the cooperative atmosphere which ultimately was created on the Bonneville project would not have happened, and the schedule would never have been kept. One or all of the stakeholders would have unilaterally or collectively "screwed up" the project.

Empathy

We have examined the concept of leadership from the initial perspective of the leader. Of course, leaders can be successful in the short term by taking command of the situation with a baseball bat and applying several forceful blows to vital areas. However, true leaders are a bit more subtle—able to identify with those under their command and take actions which are not only in their own interest, but that of those subjects. With this in mind, let us examine the next major requirement of successful partnering.

In public works contracts, there is a high probability that your contractual counterpart (adversary) has some of the same feelings as you with respect to the problem-resolution process. If you can agree on nothing else, you might agree that he or she thinks you are as big a boob as you think he or she is. The ability to feel and understand the position of another individual because you have the same or similar feelings is empathy. Empathy and understanding are in some ways synonymous. If you understand your own position and can empathize with that of your counterpart, you can feel comfortable that you understand your counterpart. To deny your counterpart's position if you truly empathize with them is to invalidate your own feelings. If you comprehend the issues from both your own position and your counterpart's, you will find that you are on the way to conflict management. Where each party recognizes each other's positions, there is a foundation for agreement. Understanding this principle is fundamental to understanding partnering.

At Bonneville the common goals of the stakeholders were always emphasized. Shortly after the "come to Jesus" meeting between KAJ and DBM, the first quarterly follow-up meeting to review the progress of the project partnership was held. Lloyd Smith of KAJ informed Michael Murphy of DBM that his presence, though not mandatory, would be greatly appreciated. It was also suggested that DBM's counsel attend. As it turned out, DBM's counsel could not attend, so Mr. Murphy went by himself. The absence was fortuitous—Mr. Murphy had to speak for himself.

The meeting began at 7 a.m. in a hotel well away from the job site. Mr. Murphy chose a chair that literally placed his back against the farthest wall of the conference hall. There were about 200 people in attendance. Mr. Murphy came with an open mind—sort of: He knew what he wanted to hear, but he doubted that he would hear it. Attempting to fade into the woodwork was an appropriate posture.

At first, the meeting resembled a junior high school dance. Where possible, each little clique sat together, with at least one empty seat between them and the next group.

Col. Cowan opened with a refresher on partnering. As he spoke, something infectious began to spread around the room. Just before the first break, the colonel mentioned something about problems with DBM's tiebacks, and asked Mr. Murphy to introduce himself. Murphy stood up and said, in a manner reminiscent of a 12-step conversion, "Hi, my name is Michael, I drill tiebacks, and I am the only #.@.* here today without an attorney." The comment was met with raucous laughter. The ice had been broken. After a break, Mr. Murphy returned and sat about midway down the aisle in an empty seat between the general contractor's quality control engineer and the Corps of Engineer's quality assurance engineer. Perhaps from this vantage point he could hear a little better.

The cycle of presentations, work sessions, and full group sessions continued throughout the day. After every break Mr. Murphy found himself sitting a little closer to the front; it is always easier to hear as you get closer to the speaker. Then the time came for an executive meeting including the brass from the Corps and KAJ. Colonel Cowan suggested that it would be appropriate to include the tieback subcontractor, since he had now moved so close to the front he was already sitting at the executive table. And so it came to pass. The bait was in the water and the reluctant subcontractor was about to have the hook set by the master fisherman, Charles Cowan.

The meeting ended about 5 p.m. with a no-host bar reception and hors d'oeuvres. Mr. Murphy and Col. Cowan spent the next several hours talking. Here was the district commander, who had in excess of $600 million in construction contracts in various stages of completion, talking to one little $4 million subcontractor. They were not talking about construction, but about life, families, likes, dislikes, and a variety of other subjects. They were discovering their commonality. They were learning to empathize.

At 9:30 that evening, as the two men stood in the parking lot of the hotel, Col. Cowan said, "Don't worry Michael. We will treat you fairly." He extended his hand. As they shook hands, Mr. Murphy knew that Charles Cowan had just put the handshake back in construction.

Common ground

Are you working more and enjoying it less? The reason for this inquiry is uncomplicated. Resolving disputes requires time and energy. This expenditure is usually made in a less-than-amicable atmosphere. If your answer to the question above is yes, why not try and do something about the situation? That is, why not introduce an element of fun or at a minimum reduce the stress associated with the confrontation related to disputes? Why not solve problems immediately, rather than let them fester and grow into disputes?

Consider that the leaders you consider to be your adversaries do not relish the combative and destructive methods usually employed to resolve disputes. Is there a hint of common ground here? Could there be some empathy? Are we not bright enough to figure out that if nobody likes what is going on, then we should change it? The answers are obvious.

We have established that stakeholders share common ground. With the potential combatants attacking a common problem without attacking each other, can a solution be far behind? Partnering provides a process and a forum which can lead to a constructive, not a destructive approach to problem solving.

On the Bonneville Lock project this simple logic was used many times. During some preliminary discussions on a pricing structure to compensate DBM for a government-owned problem, William Obley of the Corps of Engineers showed up at DBM's field office several mornings per week to discuss the problem. Though DBM was honored to receive such detailed attention to the problem, the discussions eventually became redundant. They drifted off of the subject at hand and onto other topics. One particular conversation between Obley and Murphy focused on the issue of empathy. That conversation revolved around the frustrations that owners have with contractors and vice versa. They decided that there needed to be a legal, if not socially acceptable, means to vent those frustrations. They decided that if they could just hit something and make some noise when the frustrations arose, their work would be much more satisfying.

With that cue, Murphy fabricated "The Gong"* and the accompanying poem and presented it to the Corps of Engineers several days later.

Ode to Obley

In the course of an average day,
When you allow your mind to stray,
To thoughts which are obviously wrong,
You should come out and bang on the gong.

You should never bother to fight,
Since the contractor is always right.
And to move your frustrations along,
You should come out and bang on the gong.

It's people are really quite good.
But "The Government" is misunderstood.
So don't question the contractor's song
Just come out and bang on the gong.

---

*The Gong, in slightly reduced scale, was reminiscent of one of those large brass instruments that a bunch of Buddhist monks would hit with a 50-ft teak log.

When you accept the previous truth
With a gallon of gin and vermouth
And a beauty in a silk sarong
You won't need to bang on the gong.

In the course of an average day,
When you allow your mind to stray,
To thoughts which are obviously wrong,
You should come out and bang on the gong.

You should never bother to fight,
Since the Government is always right.
And to move your frustrations along,
You should come out and bang on the gong.

Its people are really quite good.
But "The Contractor" is misunderstood.
So don't question the Government's song
Just come out and bang on the gong.

When you accept the previous truth
With a gallon of gin and vermouth
And a beauty in a silk sarong
You won't need to bang on the gong.

You can see that the song is the same.
Despite the organization's name.
There is really not much else to do,
But walk a mile in each other's shoes.

The preceding is offered to illustrate that both parties to the contract were frustrated. They had something in common and that commonality became the foundation for the solution to many problems. It also provided the opportunity to interject an element of "fun" into the serious undertaking of completing a quality project on time, under budget. One can hardly argue that there is commonality in people wanting to have some fun in their life and work.

Equity

It is not a major step from empathy to equity. For agreement to occur it is critical that the agreement be equitable. When equity is achieved, no single party wins or loses—the "right" solution is attained. And both parties feel good about it.

Equity in the context of partnering is in some ways synonymous with the common goal. In a typical construction project, equity is attained when the owner receives a quality project and the contractor realizes a reasonable profit. The third common goal is time: The owner wants early, beneficial use, and the contractor wants to maximize turnover on the resources committed to the project. If both participants are in the game for the long term, all these goals must be met. Without profits the contractor cannot deal on a long-term basis with the owner. Without quality work the owner will not deal again with the contractor. When time overruns occur, both parties suffer. When equity is achieved, the relationship flourishes, for all parties enjoy a win–win situation.

The concept of equity brings to mind one particularly difficult pricing negotiation between DBM and the Corps of Engineers. The contract modification included certain elements of inefficiency incurred by the contractor. The inefficiencies occurred due to weather, extensive overtime, and multiple-shift operations. Discussions of inefficiencies are always difficult. There is general agreement that they exist, but measuring them is another thing. The variables on each site make most studies on the subject little more than guidelines or frameworks for discussion.

This particular negotiation lasted from early in the morning to late in the afternoon with a minimal lunch break. It was grueling. By the end of the day an agreement had been reached and the contract modification was to be prepared accordingly. While leaving the conference room, Tracy Martian, the Corps of Engineers representative in the negotiation, looked rather despondent. When asked why, he informed Mr. Murphy that there were several items to which he had agreed in the negotiation but which he did not feel he had sufficient information to back up. He made clear that the modification would stand despite his reservations. After considering the situation, Mr. Murphy informed Mr. Martian that he did not expect to be paid until he (Martian) had the time he needed to assemble the facts proving the planned payment equitable. Mr. Murphy also informed Mr. Martian that if it meant a reduction in the revenue to DBM, DBM would reopen the negotiation. Mr. Martian appreciated the gesture. He did the necessary research. The modification was written as originally negotiated.* An equitable adjustment was attained. More important, trust between the parties grew.

Trust

Trust is something we all understand but might find hard to practice. There are at least three approaches to creating a trusting relationship.

---

*The Corps of Engineers later returned the favor to DBM.

Assume for a moment that there is no or minimal trust between two parties. Trust may have deteriorated over time, or it may have never existed. In this case the approach is to build it. Actions speak louder than words. In these circumstances it would be unnatural to proceed without caution. You need to start somewhere, however—as leader, make the largest and most significant gesture you can. Exceed, if only slightly, your comfort level. Take some risk, trust a little, but do not give away the farm. Work with your counterpart to find some mutually acceptable level of "discomfort" with respect to trust. As the process allows and trust builds, venture out a little further each time, always slightly exceeding your comfort level but never to a position where you assume an excessive amount of risk. The assumption of too much risk will make you hypersensitive to any perception that trust is being betrayed. In this circumstance you run the risk of taking several giant steps back, creating an even more difficult task for the future. But if you stay strictly at or under your comfort level, you will never allow trust to grow.

The second approach to building trust is to start with the commitment to trust and back off that position only when and if the trust is somehow betrayed. This method of trust recognition is easiest if there is an ongoing positive relationship or at a minimum no negative history between the players. If the players both view the failure of the trust relationship as being abhorrent or totally unacceptable, the system can work famously.

A third technique can be useful when the bargaining positions are so greatly different that the party in the lesser position has no other choice. It can also be applied when the relationships are so bad that there are few other choices. In either case, to put it in the vernacular, "If you ain't got nothing, you got nothing to lose." This might seem inconsistent with the "don't sell the farm" philosophy, but it takes guts and leadership to know when to make this bold commitment.

There were examples of all three methods of fostering trust at Bonneville. One in particular comes to mind.

Fairly early in the project it was discovered that five tiebacks on the diaphragm wall had been installed at an improper, horizontal angle to the wall. The angle of intersection was important, particularly because it affects the tieback's load-carrying capacity. Also, since the tiebacks were over 200 ft long with only a minimal horizontal separation, there was the danger of intersecting other tiebacks if DBM placed one with only a slight error in horizontal alignment. There were several causes of the problem, but the bottom line was that DBM had erred. To correct the problem with the five wrongly placed tiebacks would cost DBM about $50,000.

The schedule implications were far more serious. Installation of the tiebacks was on the critical path of the project's CPM schedule, which was already showing negative float. Correcting this problem might delay the project by two weeks; as a result of which the in-water work window would be missed and project completion would be delayed for a year.

Michael Murphy was cordially invited to another meeting with the Corps of Engineers to discuss the problem and possible solutions. Picture a rather despondent individual sauntering (he was in no particular hurry) to the conference room at the Corps of Engineers Resident Office. That conference room was filled with no fewer than 25 Corps staff engineers, geologists, and contract specialists, each armed with a three-ring binder containing the appropriate material with which to crucify the contractor.

As Murphy dawdled his way to the meeting, kicking rocks and, like a kid taking home a bad report card, finding numerous distractions, he dreamed up a myriad of excuses. Finally he settled on one and proceeded into the lion's den: "Ladies and gentlemen, I come hat in hand. I screwed up. I need help."

There was a brief moment of disbelief. Nothing was said. Then everyone closed their three-ring binders and shoved them aside. There would be no crucifixion this day. One of the geotechnical engineers said, "What can we do to help?" Within minutes a course of action was set into motion to salvage the tiebacks that could be salvaged, reengineer the wailer (the beam on the wall face that links a row of tiebacks), and reinstall the tiebacks that could not be salvaged. The direct cost to DBM was under $10,000. The work was accomplished so as not to extend the contract duration. A problem was solved. A dispute was avoided. The project was put first.

Charles Cowan told Michael Murphy that he would be treated fairly. Charles Galloway, the resident engineer for the Corps of Engineers, told DBM, "The contracting officer is fair." Mr. Murphy trusted them. His trust was not misplaced. These were honorable people. The project was the beneficiary.

Honoring the spirit

The concept of honor is related directly to trust. Honor and manifestations of honor are paramount in building or maintaining trust. An honorable person is able to admit mistakes and take responsibility, and will not take advantage of someone else's mistakes. Apply honor to both sides of a problem, and an equitable solution will result. Ignore honor, and problems will quickly become disputes.

The Bonneville project was technically difficult, and it required the commitment of massive capital and human resources. More equipment could be purchased, more engineers could be hired, more materials could be shipped, more journeymen could be added to complete the work. The element which could not be sourced from a union hall, supplier, or equipment yard was the honor of the stakeholders. No single anecdote will be recalled to emphasize this element. As you can see from the experiences cited, honor was an element of each. The Bonneville Lock project was blessed with honorable people. Its success was due in large part to their integrity.

In the partnering environment, trust begets trust. Unfortunately, honor cannot be built or created for you by others. It is up to each of us as individuals to act in an honorable fashion. Pause for a moment. Go find a mirror. Look deeply in the eyes of the person in that mirror. Is that person honorable?

The spirit of the relationship, contract, or specification—its purpose or intent—is a most intriguing aspect of partnering. As leaders we must examine the intent of any agreement and assure that we achieve the intent. Typically, the more words we add, the greater is the chance for confusion to result from contradictions. *An extremely powerful tool is available to foster a successful partnering arrangement. That tool is to honor the spirit over the letter of the agreement.* This gives the players the *flexibility* to find equity while maintaining the integrity or intent of the agreement and specifications. If a potential solution to a problem does not violate the intent of the agreement, why should the letter of the agreement be enforced to the point of changing a problem into a dispute? Clearly, it should not.

Honoring the spirit is most useful when completed work does not conform to what was expected. The unexpected result could arise from contractor error or owner error in the preparation of the contract documents. To minimize chances of not achieving the agreed-upon goal of a quality project, on time, and within budget, the solution should emphasize the intended purpose of the work rather than what might or might not have been shown in the contract drawings. This will minimize the cost in time and money to whomever is responsible for the error. The responsible party must, of course, in an honorable fashion, admit the error and be accountable for it.

On the Bonneville project there were numerous examples of the application of this principle. During the upstream excavation of the lock, it was applied on several occasions during installation of the tiebacks. The function of the tiebacks was to transfer load from a vertical wall of soil at the excavation face to anchors some 200 ft behind that face. In very simple terms, the tiebacks were designed to keep the Union Pacific Railroad on the tracks and out of the excavation. The soil in which these anchors were constructed was a remnant of an ancient, massive landslide—a mishmash of boulders, clay, sand, gravel, cobbles, water, and bedrock which made installation of tiebacks quite difficult.

A certain percentage of tiebacks usually fail to perform as designed. However, a failed tieback normally has some load capacity. A typical specification will provide that the maximum load that the nonperforming tieback will hold (as determined by test) is divided by 2, with the remaining design load and a factor of safety to be picked up by adding a new tieback. Such a specification existed at Bonneville. To follow the strict interpretation of that specification at Bonneville could have meant not making the schedule—partial failure of a tieback could have cost the project two weeks of critical-path time.

The project team had identified this potential problem early in the project.* Before any problem arose, a contingency plan was developed to deal with the scheduling impacts of a partially failed tieback. The plan included identifying critical areas where partial failures could not be tolerated, areas where they could be, means of increasing load transfer by using other existing tiebacks to make up for tieback capacity inadequacies, and keeping "spare" generic tieback tendons on site to reduce procurement time.

The project had the luxury of sophisticated electronic monitoring equipment which allowed real-time evaluation of the performance of the retaining wall. Since the intent of the wall was to keep the railroad out of the excavation, and the intent of the contract was to provide the river users early beneficial use of the lock, the concept of honoring the spirit was applied to develop a contingency plan. As long as there was no sacrifice of the integrity of the wall and the instrumentation showed no movement in the structure, the schedule could win the "schedule versus specifications" battle even if one noncritical (in an engineering sense) tieback did not meet design expectations.

The contingency plan proved to be invaluable in maintaining the schedule when several failures occurred on the upstream cofferdam wall. In this particular case, most of the tiebacks were critical. They were therefore replaced. The stock or generic tendons, already on site, were used with some minor modifications which might have been disallowed by a literal interpretation of the specifications. The stakeholders' foresight saved approximately one week for every failed tieback that was not replaced. There were several minor changes to the locations and loads of other tiebacks, which allowed the excavation to proceed with minimum delay. The end product did not look exactly like all of the straight lines on the design drawings, but the

---

*It should be noted that the Corps of Engineers and KAJ agreed as part of their partnering charter to manage the schedule jointly.

lock opened on time, and safety was not compromised. You can see how this approach is preferable to the literal interpretation of the specification, which would have required that a partially failed tieback be replaced.

The Common Thread

A very simple thing happens when partnering principles are applied. A team is built within each organization as well as a team of organizations. The reward to the individual players is measured by the success of the team. The success of the team is measured by the attainment of the project goals.

As you can see from the examples above, success of the project was put before the individual needs of the stakeholders. When this transpired a most interesting thing happened: The individual needs, too, were met, arguably far in excess of expectations.

The partnering charter at Bonneville included quantified goals plus more subjective, unquantified ones. The use of measurable goals helped the team solve problems through compromise and discussion rather than acrimonious debate. Since the targets were determined by the parties, failure in meeting a project goal could only be viewed as personal failure. The team was not made up of individuals who relished the thought of personal failure, so they worked hard to see that issues were settled and disputes avoided.

Looking back, it is clear that the project involved many challenges, but in the end there were no disputes.

Partnering in Adolescence

Partnering works in part because virtually everyone is fed up with the alternative. As Charles Cowan has often said, "By the time a dispute gets to me it is at least three years old. More than likely those that were responsible are long gone. I then have two alternatives; one is bad, the other worse!" The carrot in the partnering formula is a reduction of negative energy expended in the work environment. Partnering provides an increased chance for success.

If you only want the good news, do not read the next several paragraphs. Partnering sometimes fails. Failure to adhere to or practice any of the key concepts discussed above will make partnering more difficult.

Partnering is initiated from the top down, but it is practiced from the bottom up. If there is a lack of commitment at the highest level of an organization, there is little chance that partnering will succeed. Many studies show that the lower you go in the project organization,

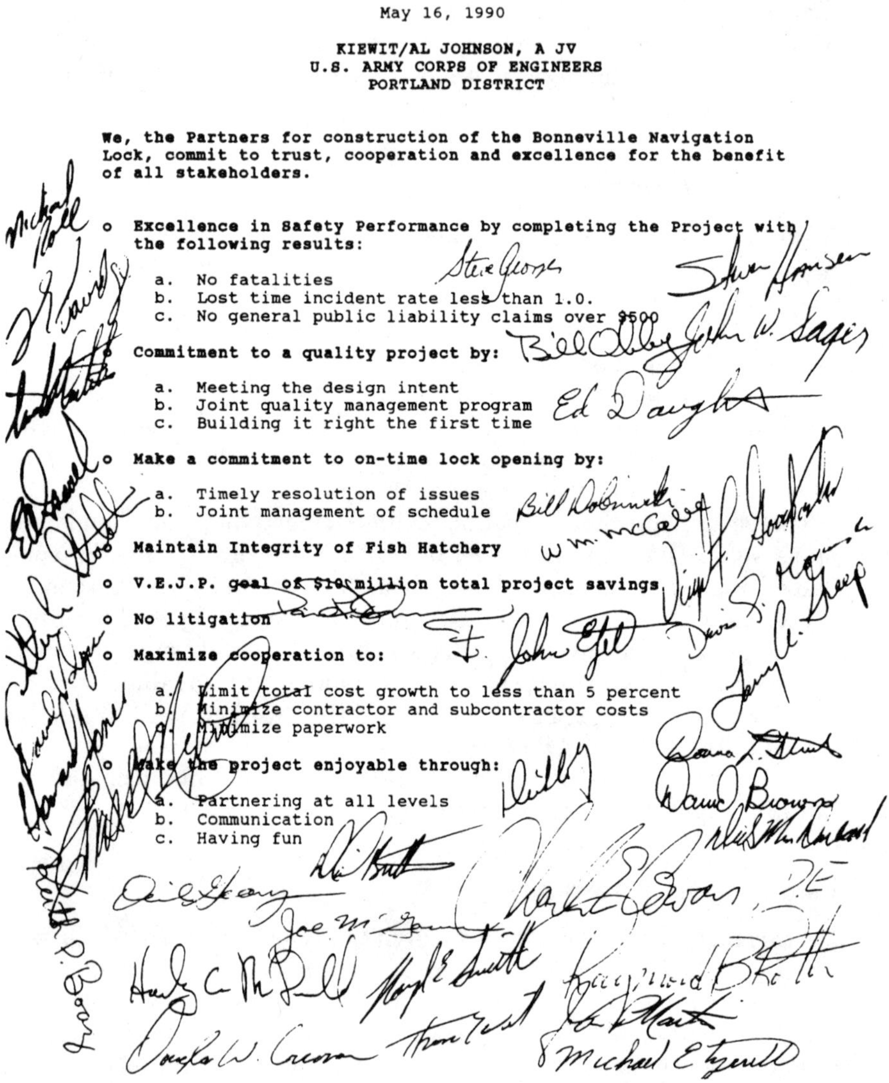

May 16, 1990

**KIEWIT/AL JOHNSON, A JV
U.S. ARMY CORPS OF ENGINEERS
PORTLAND DISTRICT**

We, the Partners for construction of the Bonneville Navigation Lock, commit to trust, cooperation and excellence for the benefit of all stakeholders.

o Excellence in Safety Performance by completing the Project with the following results:
   a. No fatalities
   b. Lost time incident rate less than 1.0.
   c. No general public liability claims over $500

Commitment to a quality project by:
   a. Meeting the design intent
   b. Joint quality management program
   c. Building it right the first time

o Make a commitment to on-time lock opening by:
   a. Timely resolution of issues
   b. Joint management of schedule

o Maintain Integrity of Fish Hatchery

o V.E.J.P. goal of $10 million total project savings

o No litigation

o Maximize cooperation to:
   a. Limit total cost growth to less than 5 percent
   b. Minimize contractor and subcontractor costs
   c. Minimize paperwork

o Make the project enjoyable through:
   a. Partnering at all levels
   b. Communication
   c. Having fun

Figure 3.1 Partnering charter signed by dozens of workers on the Bonneville Lock project on the Columbia River.

the less likely it is that a given person will have participated in the partnering retreat. Also, lower-level people do not see first-hand the pain that disputes and lawsuits cause.

Earlier it was said that partnering takes work; remember this. There is far more to it than going to a facilitated retreat which ends in a group hug. The bliss of the honeymoon soon wears off. The reality of the marriage then begins. For example, *regular meetings* to monitor the partnership must be held. *The charter must be a living tool,* not just a piece of paper hidden in a drawer. When problems arise, the partners must control the tendency to retreat to safe haven, the known quantity of combative mode. As in a marriage, failure to do the work of partnering will significantly increase the chance of the partnership failing.

Another important contributor to the failure of a partnership is using a bad facilitator. Construction partnering requires a facilitator(s) who understand(s) the need to build lasting and positive relationships between the stakeholders but who also has an understanding of the nature of the disputes the stakeholders are trying to avoid.*

Heed this advice if you wish partnering to succeed:

- Avoid any partnership which does not have a written and very specific problem-resolution scheme.
- Do not set your expectations, based on all of the hype for partnering, so high that you abandon future attempts because your expectations were not met the first time.
- Do not consider a partnering charter or mission statement a formal contract.
- Measure your results.
- Allow a sense of fun and humor to be part of the partnership.
- Include your trade labor as stakeholders.
- Work hard at the partnership.

Innovation—The Corn Story

One last story. The Shakers, virtually gone now, were a religious community who lived in celibate communal villages, primarily in New England and the Mid-Atlantic states. Men and women were considered equal. Each sex had specific duties with respect to the commune. They were housed in separate quarters. The villages were governed by elders and elderesses with equal authority. The sexes did not intermingle much except at meals and during worship. They found great joy in simple form and functional design. The Shakers gained their moniker from a rather animated dance they performed during certain worship ceremonies. During the dance they shook violently.

---

*[Editor's note: Opinions differ as to whether a construction-savvy facilitator will achieve better results. It is best to ask past clients about candidate facilitators.]

PARTNERSHIP EFFECTIVENESS EVALUATION

KIEWIT/AL JOHNSON AJV
PORTLAND DISTRICT, CORPS OF ENGINEERS

1. SAFETY ATTITUDE — 20
   - 10 – No unsafe acts or near misses observed. Employees demonstrate a positive attitude toward safety. Visitor traffic is not endangered.
   - 8 – No serious incidents observed that could result in a serious accident. Minimum public traffic complaints.
   - 6 – Occasional lost time accidents occur. Immediate corrective action is taken to prevent reoccurrence.
   - 4 – General on site attitude toward safety is poor. Improvements are required immediately.
   - 2 – Unacceptable safety program.

2. COST GROWTH — 10
   - 10 – Net cost growth projected at less than 5 percent.
   - 8 – Net cost growth less than 6 percent.
   - 6 – Net cost growth less than 8 percent.
   - 4 – Net cost growth less than 10 percent.
   - 2 – Net cost growth greater than 10 percent.

3. SCHEDULE — 20
   - 10 – Lock opening date projected to be met.
   - 5 – Delays have occurred that have the potential to significantly delay lock opening.
   - 1 – Unacceptable progress.

4. ISSUE RESOLUTION — 10
   - 10 – Proactive problem solving. No posturing.
   - 8 – Few conflicts, no unresolved problems which impact schedule, minimal posturing.
   - 6 – Occasional conflicts with some unresolved timely, some posturing via correspondence.
   - 4 – Significant posturing, defensive positioning, numerous problems and conflicts.
   - 2 – Unacceptable relationship.

Figure 3.2 Evaluation form used periodically to score the partnering effort at Bonneville Lock.

PARTNERSHIP EFFECTIVENESS EVALUATION

KIEWIT/AL JOHNSON AJV
PORTLAND DISTRICT, CORPS OF ENGINEERS

5. COMMUNICATION 10 10 – Effective.

   5 – Needs improvement.

   1 – Unacceptable level of communication.

6. VEJP SAVINGS 10 10 – Vigorous utilization of VE program and projected savings of $10 million.

   8 – Vigorous utilization of VE program.

   6 – Above average utilization of VE program.

   4 – Average interest in VE program.

   2 – Minimal interest, few VE's submitted.

7. QUALITY & DESIGN INTENT 20 10 – No rework required, quality attitude by production and QC/QA personnel.

   8 – Minimal rework and conflict.

   6 – Some rework required, occasional conflict over requirements and design intent.

   4 – Extensive rework required, frequent conflict over quality. Design integrity may be jeopardized.

   2 – Unacceptable level of quality.

Figure 3.2 *(Continued)* Evaluation form used periodically to score the partnering effort at Bonneville Lock.

A story is told to visitors of the Canterbury, New Hampshire, Shaker Village. It is said that one fall, a newspaper reporter doing a story on the Shaker community was allowed to attend a worship service or meeting one evening. The men entered and left through one door, the women through another. The gathering continued for quite some time. At one point the reporter happened to glance out the window. Through the darkness he noticed several points of light moving around in a field some distance from the hall.

After the meeting, while interviewing one of the elderesses, the reporter said he understood it was common for all Shakers that were in the village to attend meetings. The elderess confirmed that this was the case. The reporter then noted that he had seen several lanterns out in the field, with the implication being that several of the Shakers were playing hooky. The elderess indicated that all of the Shakers were present that night. She went on to say that the lights the reporter had seen were the lanterns of people from neighboring farms and non-Shaker communities taking corn from the Shakers' cornfield. The reporter asked if the people had permission. The elderess, with a rather quizzical expression, said, no. The reporter then asked what the Shakers did to stop it. Did they call the sheriff? Did they chase the thieves from the field? What did they do? Again, with the same quizzical look, as if the questions made no sense to her, she answered, "Why they are our neighbors! They are hungry! What do we do? We just plant a little more corn!"

How much we have to learn. This simple answer from the Shaker elderess sums up the philosophy of not allowing problems turn into disputes. It revisits each of the key concepts discussed above. The Shakers took charge of their destiny. They solved their problem by planting a little more corn than they needed. They had empathy for their neighbors. They looked at the intent of the human community rather than the letter of the human law. They were honorable people. There was equity in their solution, for they lived their beliefs. We would call this an innovative solution. The Shaker elderess truly did not understand the questions the reporter posed. Her answers were simply part of her nature. Perhaps sometime in the future a reporter interviewing a construction project partner will ask, "What do you do when you have a dispute with the owner?" The partner will look at the reporter and say, "We just plant a little more corn!"

Epilogue

Steve Hansen, Steve George, Dave Hablichech, Donna Street, Chuck Galloway, Bill Obley, Tracy Martian, Gary Bechtel, Art Fong, Mike Moran, Old Geo, Lloyd Smith, Dick Geary, Frank Lingscheit, Jon Dunlap, Scott Apple, Chuck Cowan, Digger Powell, Doug Corcoran, Howard Jones, Dave Johnson, Chuck Palmer, Paul Ching, John Etzel, Sam Baker, Terry Constable, Bill Dobrinski, Andy Hoff.

Above is a list of names without titles or organizations. It is important to view these names without such references. These people played a very significant role in turning what could have been a disastrous project into a success. And they converted a skeptical subcontractor into a project partner. For him, partnering has become a way of life. The success of the Bonneville project for DBM cannot be measured in economic terms. In fact, it had a less than stellar impact on DBM's bottom line. The project revenue for DBM covered only the direct cost of construction. Company overhead and indirect costs related to the project were not recovered from project revenue.

The success for DBM was in the conversion to the belief that partnering works. The adoption of the partnering philosophy by DBM has resulted in significant positive economic results for the company in the long run. Successful negotiation and early payment of equitable adjustments have improved DBM's liquid position by over 50 percent. Sales have increased by 40 percent with a 10 percent increase in margins in the face of increased competition. Management now concentrates on the critical elements of the business, including job site management, rather than wasting valuable time arguing with owners and other contractors. The employees of DBM have increased the number of smiles by 98.3 percent. They are having fun!

If I may conclude this piece in a more formal first person, I would like you to consider this. After spending an hour or so staring at the list of names above and reflecting on the project and the deep impression it left on me, I was frustrated. I found myself bemoaning the inadequacy of the language, or more appropriately my lack of command of the language, to express that impression. Surely, there must be some simple means to communicate what needed to be communicated. A variety of word combinations came to mind. None seemed to fit. All of them seemed to carry an aura of fabrication.

Finally it came down to two words. These words in their simple eloquence, properly veiled from their normal rhetorical presentation, and expressed with total sincerity, will always stand on their own merit. With a vivid picture in mind of the people the list of names represents, I offer those two words in all their simple eloquence: Thank you.

Chapter 4

How One Contractor Switched Styles from Confrontational to Partnering

Clement V. Mitchell

Clement Mitchell's firm was once said to be a "claims contractor." It converted to the project partnering philosophy.

Traditionally, the owner and contractor represent two distinct organizations with separate objectives, management styles, and operating procedures, so communication is usually restricted and very formal. Generally, each party makes decisions based on its own goals and objectives, without considering the impact on the other party. This can lead to adversarial relationships that can last the life of the project and beyond.

The objective of partnering is to design for each project a problem-finding and problem-solving team composed of personnel from both parties. Partnering recognizes and honors the objectives of all parties, thus creating synergy for project success. When we partner, we talk before we write. We promote pride in workmanship by all members of the partnering team. We maintain project commitments via periodic feedback sessions, to determine whether we are meeting objectives. Our follow-up sessions are open, attended by the entire partnering team, and we are sensitive to each party's problems. A meeting is never adjourned until everyone is heard and a group commitment has been made to resolve problems.

MCI Constructors, founded in 1974, was like most successful industrial contractors during the late 1970s and early 1980s. It was familiar with its rights under the 1978 Contract Disputes Act, and made every effort to take advantage of opportunities to collect additional costs associated with design defects and contract changes. Because MCI was successful in the pursuit of requests for equitable adjustments, it became known as a claims contractor. As a result, its reputation was damaged, since owners felt that MCI was of the breed of contractors that would exploit contract terms in order to take full advantage of additional costs it could collect that were associated with design deficiencies.

The Contract Disputes Act established certain rights under federal government contracts to provide a fair, balanced, and comprehensive system of legal remedies to resolve contract claims and to ensure fair and equitable treatment to contractors and government agencies. One of the act's intents was to eliminate contingency funds that contractors generally built into their bids to cover unknown costs, including those associated with changed site conditions and design defects. The rationale was to establish a system whereby contractors would eliminate contingencies but could make claim for extra work and thereby reduce bid costs to the government, since all projects in the competitive bid process did contain such unknowns.

Unfortunately, some agencies failed to comprehend the intent and purpose of the act and were offended when contractors began exercising their rights to be compensated for additional costs, particularly since prior to implementation of the act, conflicts such as design changes and other unforeseen and unanticipated conditions were usually taken care of by contractors through their contingency allowances. I believe the silly myth that some contractors bid at or below cost to get a job, then make its profit on change orders, was created by owners who did not understand the Contract Disputes Act. No contractor in his or her right mind would really bid like that.

MCI is a heavy industrial general contractor specializing in managing and constructing complex industrial power and mechanical projects and air, water, and wastewater environmental facilities. When I took over the company in 1988, I realized that even though MCI had been incredibly successful in the pursuit of claims, they were pyrrhic victories. No one had measured the costs to the firm in damage to its reputation. This led to the reluctance of owners to work with a company that would sue them at the drop of a hat. Also, no one considered the administrative distractions associated with claims, such as the time and energy spent in depositions, briefs, and various other requirements of the litigation process. This all took away time from the company's primary tasks of pursuing new work and managing it effectively.

If you plan to engage in construction, you must anticipate controversy. Such controversy existed on the TNT Thick Liquor Facility for the Army Corps of Engineers in Radford, Virginia. This should have been a research and development (R&D) project, but the Corps decided against R&D even though the process the plant would house had never been tested. The design intent was to construct a $19.3 million facility to operate a sulfide-recovery process for treatment of TNT "thick liquor" (red water) derived from purification of TNT and to recover sodium sulfide for reuse in TNT purification.

Shortly after I took over MCI, the owner terminated this project for convenience, as a result of numerous design problems. This was nearly three years after notice to proceed; the Corps did not feel a redesign of the facility under MCI's contract would be cost effective. In fact, its estimate to complete the proposed redesign and construction of the project at that point was over $45 million, more than twice the original contract price. Federal government contracts usually contain "termination for convenience" clauses which give the owner the right to terminate all or any portion of a contract due to conditions beyond the control of the government. This clause is also intended to be used in the event that conditions arise which would prevent the contractor from proceeding with or completing the work, or for any other conditions that would prevent the government from continuing with the contract.

Prior to the termination for convenience decisions, there was an extreme adversarial relationship between the Corps and MCI Constructors' representatives. MCI made hundreds of requests for information and submitted a voluminous claim to cover the cost of the anticipated extended performance cost due to design defects. There were numerous problems with the heat balances and mass balances of the multiple-hearth furnace that was specified by the government. The Corps was reluctant to provide the requirements of the air inputs for this furnace as well as a total furnace feed rate that was mandatory so MCI could determine equipment sizing. Conflicts also arose over the sizing of the secondary air stacks for the furnaces.

The Corps' attitude was one of arrogance. Corps officials repeatedly told MCI to make the project work, while at the same time they withheld critical design criteria. The weights of the equipment that the designer had used as the basis for structural steel calculations were withheld. The Corps' only response was that nothing necessary for the building and procurement of the facility was withheld intentionally or unintentionally. It continued to hold that the contract was complete and all that was necessary to produce the facility had been provided to the contractor. This created critical problems affecting MCI's ability to coordinate and build the project. Up until the time of the termination for convenience, the Corps refused to recognize the specification deficiencies or provide the information necessary to overcome the deficiencies.

In the midst of this chaos, I met Col. Richard Sliwoski of the Corps. He had recently taken over command of the Norfolk District and was acting as the terminating contracting officer on MCI's Radford, Virginia project. Before meeting with me, Col. Sliwoski was briefed on my reputation; he later told me he was advised that MCI personnel were the devil incarnate and would fight at the drop of a hat. Just as the colonel was briefed on our reputation, I was briefed on his. I heard that he was tough but fair and that he was an independent thinker.

Although we work for different organizations, we had similar management styles. We both are dedicated professionals who believe in fair and equitable treatment for everyone with whom we associate. Neither of us tolerates the folly of fools easily, and we both respect those, like us, who have truly paid their dues and earned their stripes.

A two-day settlement meeting on the Radford project was attended by a large group of MCI and Corps representatives, including two attorneys from each side. When the discussions reached an impasse, the colonel and I met in his office for approximately 10 minutes and resolved the claim. Unfortunately, certain civilian representatives of the Corps were still adamant that we should have made the project work, despite the design defects, and went on a campaign to damage MCI's reputation. In fact, during the height of MCI's initial discovery of the design deficiencies, these people informed MCI that it intended to issue an interim unsatisfactory performance evaluation. MCI was forced to respond with a voluminous document to this threat, which we felt was intended to intimidate us into assuming redesign responsibility for the process equipment, contrary to the contractual assignment to the Corps of such responsibility. This campaign by some former Corps civilian employees continues to this day.

Shortly after receiving the termination for convenience on the Radford project, we bid a project at Fort A. P. Hill, Kentucky, for the Corps, which consisted of a very tight, 240-day and 480-day schedule to install 30 miles of underground power distribution lines, 17 miles of water lines, and 12 miles of gravity sewer within 240 days, prior to the worldwide Boy Scout Jamboree of 1989. Film director Steven Spielberg and then-Vice President George Bush attended this very high-profile event. We accomplished this Herculean task as per contract, but on the second 240-day phase we wound up back at our adversarial position on the completion of Phase II of the plant, which included a wastewater-treatment plant and various supplementary systems to that plant.

Due to delays incurred as a result of work switched from Phase I to Phase II, we submitted a claim to the contracting officer and wound up settling this claim via an alternate disputes resolution (ADR) process. On this project, we started with a partnering concept to meet the jamboree deadline, left it after the jamboree, and then returned to the concept by utilizing ADR to settle the contract.

The Fort Dix Wastewater Project

Despite the problems with the Radford and A. P. Hill projects, I remained convinced that we could work with the Corps. I looked at a project at Fort Dix, New Jersey, for the Corps' Philadelphia District. By that time I felt that if it is true that you learn from adversity, I was evolving into a very bright person. A provision of the contract for this project allowed the utilization of the partnering process if both parties agreed to participate. I looked at partnering this way: Nothing else we had tried had worked, and maybe the third time with the Corps would be a charm.

We targeted the Fort Dix–McGuire Air Force Tertiary Wastewater Treatment Plant project due to its complexity and size. It was a mid-sized project ($34 million) and included demolition of an existing plant and a crucial schedule, since the Army was under court order to meet an EPA compliance mandate. We had to pour over 16,000 yd^3 of concrete in less than one year, and operated the entire project within two years of notice to proceed.

The project consisted of the construction of 10 new structures, a state-of-the-art Bardenpho Aeration tank system, 48,000 linear feet of 48- and 24-in force main, installation of mixers, clarifiers, sludge-handling and filter equipment, and a 24-acre land application site. During the project, it was determined that McGuire Air Force Base would remain open, despite the fact that it had initially been targeted to be closed under the Base Closure Act. As a result, we were required to expand the Bardenpho system by adding a third process train while still completing the project within the specified time.

The Bardenpho system, using a series of tanks (each about 400 ft × 200 ft in area), is an advanced secondary wastewater-treatment process designed for biological nitrogen and phosphorous removal. It is a staged, activated-sludge process which uses the biochemical oxygen demand (BOD) of the wastewater for biological nitrogen removal. The five stages of this process include fermentation, first anoxic, BOD/nitrification, second anoxic, and reaeration stages.

Partnering on the Fort Dix Project

The transition to partnering was difficult, as with any other change in philosophy. As Machiavelli said over 200 years ago, "There is nothing more difficult to carry out, no more doubtful as success, no more dangerous to handle than to initiate a new order of things." Partnering has been extremely successful on the Fort Dix project due to the partnering members, the owner, the engineer, and the contractor, and their total commitment to the partnering charter we developed in our first meeting.

In this charter (Fig. 4.1), all the participants agreed to work together as a cohesive team to produce a quality project, in accordance with the contract, on time, within budget, and safely, while enabling the contractor to earn a fair profit. Members of the partnering team also agreed to:

1. Deal with each other in a fair, open, trusting, and professional manner. In that spirit, we agreed to communicate openly, resolve problems, and
2. Make decisions at the lowest possible level.
3. We agreed to maintain a professional atmosphere, talk before we wrote, and to promote pride in workmanship by all members of the partnering team.
4. We maintained these commitments by having periodic feedback sessions to determine whether we were meeting our objectives. Our follow-up meetings were open sessions that were attended by the entire partnering team, whose members were sensitive to each others' problems.
5. A meeting was never adjourned until everyone had been heard and a group commitment made to resolve any problem areas.

We have performed over $8 million in change orders at the request of the Corps of Engineers, and have added only 15 days to the schedule.

It appears that the Corps is happy with our work: In 1994, I was awarded the Corps' prestigious Commander's Award for MCI's quality and timely construction of the joint Fort Dix and McGuire Air Force Base Tertiary Wastewater Treatment Plant and for my leadership and commitment to the partnering concept.

**PARTNERSHIP CHARTER BETWEEN
MCI CONSTRUCTORS, INC. AND THE ARMY CORPS OF ENGINEERS**

COMMUNICATION

GOALS	ACHIEVEMENTS
1. Communicate problems openly and as early as possible.	1. Communication was very open on the project. Matters were discussed freely without posturing.
2. Resolve problems and make decisions at the lowest possible level in a timely manner.	2. Most difficulties were resolved at the project management level with the site project managers and engineers.
3. Maintain professional atmosphere of mutual respect and resolve personal conflicts immediately.	3. A high degree of respect and professional demeanor was demonstrated throughout the job by all parties.
4. Talk before we write.	4. There was virtually no written documentation or posturing on the project.
5. Develop periodic feedback program on the partnership's communications.	5. We conducted periodic partnering workshops, that were attended by all partners, to evaluate performance.

Figure 4.1 Fort Dix (New Jersey) Wastewater Treatment Plant project: partnering goals and accomplishments.

PARTNERSHIP CHARTER BETWEEN
MCI CONSTRUCTORS, INC. AND THE ARMY CORPS OF ENGINEERS

PERFORMANCE

GOALS	ACHIEVEMENTS
1. Produce a quality product the first time through an effective and committed quality management program (QA & QC).	1. QC programs by the parties were extremely effective and one of the reasons we are able to complete the job in a timely manner.
2. Allow partnership contractors to earn a fair and reasonable profit.	2. Both parties were satisfied with the budget results.
3. Complete project ahead or on schedule (avoid delays).	3. $8 million in changes were added to this $34 million project with an 850 day duration, while keeping time growth within 15 days to satisfy a court order for compliance.
4. Perform work in a safe manner minimizing lost time injuries (no fatalities).	4. MCI received two safety awards on the project.
5. Promote pride in workmanship by all members of the partnering team.	5. MCI received the coveted Commander's Award on the project.
6. Minimize formal disputes (no litigation).	6. No attorneys were hired and no litigation was filed.
7. Ensure successful project start-up w/smooth transition to client.	7. The start-up went well and the project is currently operated on a daily basis for its intended purposes.

Figure 4.1 (*Continued*) Fort Dix (New Jersey) Wastewater Treatment Plant project: partnering goals and accomplishments.

The Colonel-and-Clem Show

As a result of the success on the Fort Dix project and our overcoming of the adversarial relationship between the Corps and MCI, Col. Sliwoski (now Chief of Programs and Planning for the U.S. Army's Environmental Program) and I have given speeches on partnering at several meetings. Figure 4.2 is a quote of the ancient Greek Homer, and one of my favorite partnering statements.

In our presentation, entitled "A Partnering Odyssey" (or "The Colonel-and-Clem Show," as my wife and daughters call it), we explain the partnering process and trace the association between the Corps and MCI Constructors, from the time it was an adversarial relationship to the outstanding one we now enjoy.

I have also been a partnering facilitator for the Corps' Baltimore District, and with Parsons Management Consultants for the Metropolitan Washington Airports Authority project.

MCI places a heavy emphasis on efficient project management and on prompt identification and resolution of disputes while attempting to avoid litigation at any cost. We have found that the most successful way to achieve these goals is to utilize the partnering process. Partnering advances the goals of both the owner and the contractor to complete a project on time and within budget and generally eliminates unnecessary litigation.

This alternative management process attempts to resolve the problem of several organizations being involved in each construction project. The objective of partnering is to design for each project an effective problem-finding and problem-solving management team composed of personnel from both parties, thereby creating a single culture with one set of goals and objectives.

Partnering's primary advantage is that it recognizes and honors the objectives of all parties, thereby creating synergy for project success:

- The owner has goals of completing a quality project safely, on time, and within budget.
- The contractor wants a quality project as quickly as possible, at a profit.

There is commonality among these goals. The partnering process provides a vehicle for enhancing and recognizing mutual goals and encourages a cooperative working environment.

Traditionally, construction projects present a number of challenges. The owner and the contractor represent two distinct organizations with separate objectives, management styles, and operating procedures. Consequently, communication is usually restricted and very

> May the gods give you all the things which your heart desires... for there is nothing better than this — when a (contractor) and (owner) keep a project in oneness of mind, a great woe to thier enemies and joy to their friends, and win high renown.
>
> *The Odyssey*, book VI, line 180

Figure 4.2 Contractor Clement Mitchell and the Army's Col. Richard Sliwoski have given several talks on partnering. This is one of Mitchell's favorite slides used in illustrating the lecture.

formal. Generally, each party makes decisions based on its own goals and objectives, without considering the impact on the other party. The separate operational processes followed by each party often result in adversary relationships that can last the life of the project.

The frequent result is construction delays, difficulties in resolving disputes, cost overruns, litigation, and a win/lose climate. I feel that partnering is the most successful and cost-effective way to build a project, and I have introduced and proposed partnering on all our projects.

MCI has a very successful relationship with the Metropolitan Washington Airports Authority and its agent, Parsons Management Group, as a result of the partnering process. MCI is building a complex boiler/chiller plant with tight schedules at Washington's National Airport. It is crucial that the schedule be met on this operating airport that accommodates over 45,000 air travelers a day.

The partnering process has already paid dividends for all parties by resolving design conflicts promptly and effectively. This immediate response to these problems will help us meet the project deadlines.

We also have an informal partnering process on a wastewater-treatment facility with the Beltsville Agricultural Research Center in Beltsville, Maryland, and with Southeastern University for its project at the Continuous Electron Beam Accelerator Facility in Newport News, Virginia.

Since partnering has been very successful for our organization and has helped us turn around our reputation as well as given us a new, exciting, and profitable way to perform work, I endorse the process without any reservations.

Partnering Is No Panacea

Partnering does have limitations and disadvantages. Regardless of how successfully the procedure works during the beginning or in the middle of a job, if an owner or design firm decides to discontinue participation in the process, a contractor could be left battling disputes with little or no documentation. The partnering process virtually eliminates standard letters and requests for information and encourages open, verbal communication. Consequently, there is very little correspondence to support a party's position in a dispute.

To ensure that the partnering process continues for the entire length of the project, the senior managers of each organization involved must be intimately involved in the process. They must maintain the courage to overcome obstacles and control the partnering process by maintaining a commitment to the goals and charter of the partnering team.

One other concern of contractors is that, on partnered jobs, they are generally expected to perform extra work for no charge because they have become friends during the partnering process. Contractors are in business to make a profit and cannot work for free. Likewise, a contractor cannot attempt to take advantage of the partnering concept by requesting monies for anything that closely resembles additional work. There has to be a middle ground. To obtain this, both parties have to develop and maintain trust in each other and stand behind their commitments.

Chapter

5

Can Architects Return to the Construction Site via Partnering?

James H. Bradburn

Why is there so much enthusiasm for project partnering? James Bradburn, of the Denver architect C. W. Fentress J. H. Bradburn and Associates, tells how it transformed the construction of Washington state's Natural Resources Agencies Building in Olympia.

The project had reached a point where all parties—the owner, the constructor, and the architect—felt disenfranchised from the process and powerless. The constructor was claiming he was incurring extra costs, for which he wanted reimbursement. The owner felt that the economic value of the project was declining due to threatened claims. The architects felt powerless to force the constructor to respect the design intent.

Then they tried partnering. The results were spectacular. After the initial sessions and buy-ins, communications returned, solutions rather than defensive posturing were promoted, and the sticky issues of cost overruns, budget, schedule, and maintaining design intent were solved to mutual satisfaction.

What Do Architects and Constructors Do?

Let's begin this chapter with some basic understanding about what architects and constructors do. I know that sounds simple, and I am sure that most architects and constructors would quickly tell me they know what they do very well, thank you. But I am always amazed at the misunderstandings that exist about their respective activities, by both the practitioners themselves and, most important, the owners and the public. So let's pose the question again: What *do* architects and constructors do?

Architects start with a blank sheet. Their instructions are sometimes given as specific requirements by owners who are clear about their intent and concise in their delivery of information. Sometimes the owner's instructions and communications to the architect are in the form of hopes, desires, and wishes, even pictures of what the owner likes, or vagaries that require considerable skill by the architect to coalesce into three-dimensional forms. Other times the information is not specific, or is unclear, or requires considerable research of published standards. My experience with large public projects, and in particular with the federal government through the General Services Administration (GSA), has been that these owners produce such large amounts of information about the programmatic standards to which their projects are to be designed, the volume itself almost guarantees that conflicts will exist within the information given in the documents. I applaud recent initiatives within the GSA to attempt to reduce the regulations which presently require extremely tedious effort by the architect to understand what the client wants.

Architects take these verbal or written instructions (usually called the "program") from the owner and begin to create a design. The design is developed through the traditional phases of schematic design and design development by drawings, words, and sample materials—even computer-aided renditions and video walkthroughs. This entire effort, culminating in the design development documents, is for the sole purpose of confirming that the owner understands the design and agrees that the architect has met his requirements and followed his instructions.

After the design has been established and documented in the design development documents, the architect begins the construction documents. It is important to understand that at this crucial juncture in the project, the architect has completely changed his orientation. He is no longer creating a design but rather producing a set of instructions, in written and pictorial form, that will allow a hopefully reasonably able constructor first to price and then to build the design.

In my practice, and I suspect in most other architectural firms, this demarcation between designing and producing working drawings is at times blurred or even nonexistent: Design development may not be completed before the construction documents are started. The pressures of schedule and internal budgets are usually the culprits, but the consequences of not completing the design can be very damaging.

It is really very simple: If you are not clear about the design, how can you be clear about the instructions you prepare for the constructor? If you are not clear about the design, how can your design subconsultants be clear about what their work will be? In my office, we attempt to monitor and assess this passage by checking the design development documents against an office standard list of tasks that must be completed and shown on design development documents before we allow the process to continue.

Constructors, on the other hand, do not start with a blank sheet. They are given a set of instructions that describe in detail what they are to build. There is plenty of room for creativity by the constructor on *how* to build the project—i.e., how to interpret the instructions and sequence the work—but he does not have to guess about *what* he is going to build. His job is to follow the instructions given him by the architect.

This is where all the "fun" begins. Since buildings are unique and complex creations, no set of instructions as to how to construct them will ever be without errors or omissions. Remember that the instructions are created by normal human beings who have checked their work in a professional manner and have attended to all the details with due care and diligence; and yet, mistakes are made and information is still missing that, in hindsight, would certainly have helped the constructor understand the architect's intent. On top of all this potential for errors or missing information, the instructions (the construction documents) are *always* subject to the constructor's interpretation. This is true regardless of the time spent producing the documents, the expertise of the architect's office staff, or the native intelligence of the architect himself. Anybody who has ever baked a cake (a relatively simple exercise) has experienced the frustration and inedible consequences of misinterpreting the recipe.

Consequently, I believe very strongly that the architect must stay involved in the construction of the project. He must be there to assist the constructor in understanding the instructions, to fill in the omissions and to correct the errors, but most important, to interpret the intent of the instructions. The architect must be sure that such interpretations of intent keep the project within budget and schedule.

What Caused the Separation between Architects and Constructors?

Now that we know in general terms what architects and constructors do, one creating a design from a blank sheet of paper and one constructing a building from a complex set of instructions, let's turn our attention to the central issue of this chapter: the architect's commitment to lead on the construction site.

In order to lead at the site, the architect needs to be at the site, or at least involved in the construction processes—doing what is commonly called *contract administration*. Over time, there has been a measurably reduced involvement of architects in the construction process, a separation between architects and constructors. Why has this separation occurred? And why has architect involvement in the construction process been reduced? Perhaps a brief history, as shown in Fig. 5.1, will help.*

Beginning in the late eighteenth and during the nineteenth century, American society was transformed from a primarily agrarian approach to the production of goods (the farmer performed all tasks necessary to grow produce and deliver it to market) to one featuring industry that was highly specialized, relying on individuals and organizations to produce a specific product or service. The construction industry was not immune from this process, and the concept of architect as master builder, responsible for all aspects of design and construction, changed into separate entities of design, led by the architect, and construction, led by the general contractor. (I find it amusing that so much is writ-

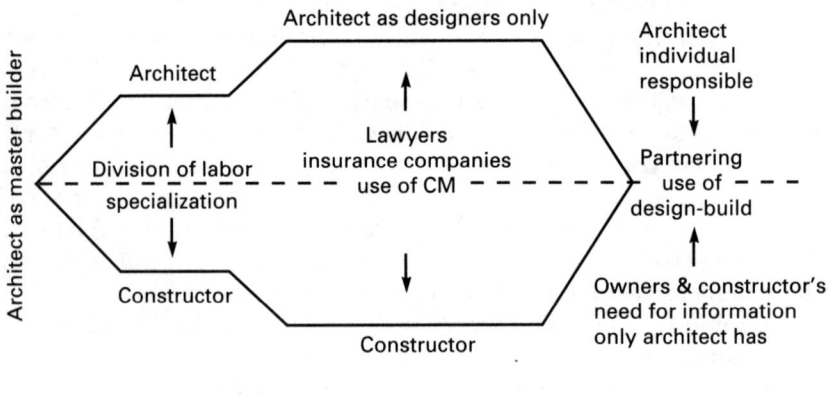

Figure 5.1 History of project design and construction. At left is shown the arrangement in earlier centuries—the age of the master builder. In the center is shown the situation in recent decades—a split between designers and contractors. A possible today and tomorrow is shown at the right—a reuniting of designers and contractors, thanks at least in part to partnering and design-build.

---

*For a more detailed discussion of political influences which weakened control of building practices by architects and enhanced the importance of constructors, see Thomas S. Hines, *Burnham of Chicago: Architect and Planner* (Oxford University Press, New York, 1974), and Roger A. Chandler, United States Domestic Architectural Marketing: History and Methods, 1909–87 (Ph.D. dissertation, University of California at Santa Barbara, 1990).

ten today about the "new" approach to project delivery called design-build, but that's a subject for another book.)

The process of separation did not stop with the advent of the twentieth century. During the past 40–50 years, the separation of the architect from the construction process has been accelerated by the rising tide of litigation, and architects' resulting withdrawal from the construction site, on the advice of their attorneys and insurance agents. Architects have in the past been consistently told to limit their involvement at the site so as not to give rise to any argument that they might have control over the construction process. For example, insurance companies that provide architects with professional liability insurance advise their clients not to use the word "approved" when checking shop drawings, but instead to use the phrase "no exceptions taken." I remember one constructor telling me that his interpretation of that phrase was "I can't be bothered." How absurd can we get, and yet what a strong signal such actions by architects send to constructors and owners about the architects' desire to be involved with construction.

Alas, I have digressed. There are other forces at work here. During the last 20 years the separation has been further aggravated by the increased competition among architects and the consequential market-driven lowering of design fees. Something has to give: Lower fees means less service offered. And with litigators and insurance providers breathing down the architects' collective necks, the architectural community as a whole, and constructors and owners in general, have come to the conclusion that by eliminating construction-phase activities, architects would lower their liability (less litigation and insurance costs) and reduce their fees (less service to render).

What Are the Results of the Separation?

This separation has results that are quite damaging to architects individually and to the profession. As part of their professional education, architects are trained to understand how things are built, how to investigate the serviceability of materials and products, and how to manage and work with people to achieve mutually satisfactory goals. When the architect is removed from the construction process, this training is not utilized and, consequently, the skills are diminished or lost. Additionally, when they are removed from the construction marketplace, architects become less familiar with construction pricing and, as a consequence, less able to represent the owner's interest of designing projects within budget.

The profession is hurt as well. The less familiar architects are with construction techniques and pricing, the less skillful at working with constructors with varied goals and, at times, cross-purposed agendas represented by all the tradesmen and suppliers, the less effective and meaningful architects become in the construction process. The profession comes to be viewed in an ever-narrowing role. The separation has further reinforced the owner's desire not to pay the architect for these services because members of the profession are viewed as only "designers," not full-service architectural professionals.

However, the owner, and most importantly the constructor still need someone to correct the errors, fill in the omissions, and work with the constructor to interpret the instructions in the contract documents. So who provides this information? Not the architect—he has essentially abdicated the responsibility or been removed from it during fee negotiations by an ill-informed client. It is my belief that this void has given rise to the use of "construction managers," a phenomenon that has emerged and blossomed into a high-growth industry cloaked in considerable confusion about what it is and what exactly this "manager" manages.

Construction management is neither construction nor management. The constructor does the construction and manages himself just fine. He does not need someone who has no real responsibility for the end result to tell him how to do his work. But the constructor still needs information from the architect. The construction manager does not have the necessary design information to complete the construction or the ability to interpret the original design intent. He must get that from the architect. The construction manager collects a large fee merely to stand in the void between architect and constructor, without risk, passing information between the two. So whence cometh this construction manager? Because the need for information and interpretation is still there, and owners realize that they need someone to represent their interest at the job site—a role that in the past was fulfilled by the architect.

Thus the results of this separation are architects with diminished skills and reduced influence in the construction process; owners and a general public who view architects as merely "designers," not professionals who understand the rigors of construction and who can represent their interest vis-à-vis the constructor; and finally, the rise of another breed of bureaucratic managers performing redundant services that are best performed by architects and constructors.

What Should Architects Do about the Separation?

Architects should reconsider their role during construction. This process of separation can be reversed only by the architect as an individual and by the profession as whole. Architects must take the initiative, and they can do this only by committing themselves to taking a leadership role on the site, to represent the intent of the design and the interests of the owner, and by demonstrating to owners that the role of the construction manager is best performed by the architect. The need of constructors for information and clarification has not diminished. (Many would say that the need for information is increasing dramatically due to the increased complexity of building systems and the degradation in the quality of construction documents. Many architects take no exception to this position—there's that phrase again!—but say that fees are too low to allow them to provide full information.)

Architects need to reassess the recent forces that have accelerated the separation. The premise that architects have less liability if they remove themselves from the construction process is groundless. If there are major problems that result in litigation or arbitration, it is almost a given that the architect will be named in the complaint. And since owners are now paying fees to construction managers (who really contribute very little to the design and construction process), why should architects not reclaim those fees in order to furnish the services that owners need and that are best provided by architects?

Architects must be proactive in anticipating construction problems and providing timely solutions to assist the constructor. They must realize that their instructions to the constructors (the construction documents) will contain errors and omissions, and their best defense is to be there to correct the errors and provide the missing information before the missing information becomes a serious issue or the building is constructed improperly. Architects should be representing the owner's interests, and they can do that only when they are professionally up to date on construction practices and are knowledgeable about current construction costs and pricing.

Finally, architects need to take a leading role in representing the intent of the design. The instructions to the constructor, like any other complicated sets of instructions, are subject to interpretation. The possibility of misinterpretation by the constructor is great, and the architect (who knows the intent of the design) is the one who can best assist the constructor in understanding that original intent. It certainly makes no sense to run the communications through a third party (such as a construction manager). We all know the story of how an event will be reinterpreted when it is communicated around a circle of people. Construction is complicated enough. Let's keep it simple!

What Has This to Do with Partnering?

Partnering is an organized method to achieve a mutually desired result by balancing the three essential forces that shape every construction project. These forces can be roughly described as follows:

1. The constructor's desire to meet budget and schedule
2. The owner's desire for a project of economic, emotional, or political value
3. The architect's desire for quality and integrity of the design intent

If you remove one of these three forces, the project becomes like a two-legged chair and begins to lose stability. Therefore, partnering, as a process of continuous dispute prevention, actually promotes the concept of a strong leadership role at the site for the architect and requires that any separation between constructor and architect be eliminated.

One of my projects where this need for balance was made obvious to all the participants was the Natural Resources Agency Building in Olympia, Washington. Constructed under a design-build contract, the project had reached a point where all parties, the owner, the constructor, and my office, felt disenfranchised from the process and powerless to affect the outcome. The constructor was claiming that extra costs due to unforeseen conditions were adversely affecting his budget and schedule. The owner felt that the economic value of the project was declining due to threatened claims by the constructor. We felt powerless to force the constructor to respect the design intent and keep in the project those aspects of design that were originally intended.

About midway through the project, things looked very bleak. Communications were breaking down, fingers were pointing, and threats emerging. Fortunately, the owner became aware of a new concept called partnering, proposed by the Corps of Engineers in Portland, Oregon, and sent a representative to investigate and determine if partnering would be appropriate and useful on our project.

The results were spectacular. After the initial sessions and buy-ins, communications returned, solutions rather than defense posturing were promoted, and the sticky issues of cost overruns, budgeting, scheduling, and maintaining design intent were solved in mutually satisfactory ways. The process was hard; it required strong commitments by all parties and reinforced the need for the architect to take a strong and useful role in solving construction-related problems. Today, the Natural Resources Agency Building is an award-winning building, viewed with pride by all the participants in the process of its design and construction.

Partnering depends on the architect being knowledgeable about construction processes and costs, so he can understand the constructor's point of view. The architect must be technically competent so that he can quickly provide detailed solutions to unforeseen conditions, errors, or omissions in the instructions. Partnering requires the architect to understand the owner's interest in an economically viable project.

Partnering also requires that the architect possess good interpersonal skills. The interest of all participants under a partnering charter can best be served by a cooperative approach to problem solving. Strong representation of interests while maintaining respect for opposing views are skills that are essential to good partnering.

Does the architect, as a strong leader on the site, now assume some of the obligations of the constructor? Absolutely not—the architect is there to assist the constructor in the construction process, not to perform the construction. He is not responsible for the constructor's means and methods, and he is not responsible for safety. However, he is an integral part of the construction process, as the originator of the design, as the interpreter of the instructions given within the guidelines of the partnering charter, and as the representative of the owner on site. The architect must regain this leadership position and close the gap that has developed over the years. The project and the owner, as the architect's customer, deserve nothing less.

Conclusion

Once one understands about what architects do and what constructors do, I believe it becomes clear that in order to complete the complicated process of design and construction, the separation between architects and constructors must be eliminated. Architects again must take a leadership role on the site and become full partners in the construction process. Partnering charters are statements of individual and corporate commitments to mutually achievable goals. To achieve successful projects, architects must be willing to commit themselves to lead on site—narrowly for their individual and professional betterment, more broadly for the owner's interest and representation, and finally for the higher good of the project and design intent.

Chapter 6

One Partnering Success Secret: Set High Goals

Jim Gans, Gretchen Gagel McComb, and Ed Wambsganss

One of the more successful partnered jobs was the recent $35 million addition to a wastewater-treatment plant in Clark County, Nevada. The job was finished with no OSHA-recordable injuries, and 12 months ahead of the 35-month schedule. The competence of general contractor Western Summit (Denver) and the use of partnering were key, as was the setting of very aggressive yet attainable goals.

Engineer Black & Veatch (Kansas City) agreed to a process that would allow some communication between the parties prior to a rejection of a contractor's shop drawings. Thus the contractor and the engineer could negotiate in the name of optimizing cost, quality, and schedule.

Background

Project partnering is successful to varying degrees, and the variance often stems from the overriding culture of the organizations involved. Some organizations repeatedly experience a lack of success in partnering. This failure manifests itself in many forms, including managers who are not committed to the team atmosphere, lack of trust among participants, and high levels of frustration. Unsuccessful partnering often results in financial failures, such as claims, cost overruns, and schedule delays.

One explanation for an increased level of effectiveness with partnering comes from those organizations that are implementing some type of continuous improvement program such as total quality management (TQM). Although most people feel that they have always been "cooperative," partnering is about doing business in a completely different manner, and realizing the benefits of change through reduced costs and increased efficiency and productivity. Partners are encouraged to break through their paradigms and view their business in an entirely different light. TQM reinforces the same message. Employees are encouraged to do business differently, to look at processes differently, and to redesign the way they work. They become more comfortable with the dreaded word "change," and learn to embrace new ideas because they have "ownership" of those ideas.

In order to clarify project partnering successes and failures, this chapter presents a case study of a highly successful project.

You have probably seen bumper stickers asking where is some unknown town in the United States. Laughlin is one of those towns. Tucked into the southernmost tip of Nevada, 90 miles south of Las Vegas, Laughlin was just a sleepy fishing spot along the Colorado River in the early 1980s. It boasted a population of fewer than 100 until land in state hands was offered for development.

By the mid-1980s it was a veritable boomtown, which could be compared to the growth of the western mining towns of the 1800s. Almost overnight, major gaming casinos sprang up, and it did not take long before Laughlin casinos were making a significant contribution to Nevada's economy. Hotel chains such as Hilton, Harrah's, and Ramada built major facilities in Laughlin; the Hilton operates a 2000-room hotel/casino. In all, 10,000 hotel rooms were built to meet the demands of the burgeoning tourist industry. Laughlin's population increased 10-fold by 1985, and by 1994 was approaching a permanent population of 10,000.

Obviously, Laughlin's hot, dry climate is not a deterrent. One reason is that the Colorado River and Lake Mojave Recreational Area are close at hand. In fact, the climate seems to attract tourists in the winter because temperatures are in the 70s when other parts of the country are snowed in and experiencing freezing temperatures.

The ability of public and private utility companies to keep up with this phenomenal growth was tested. It seemed as if the normal business posture was playing catchup. No water or sewer facilities existed in Laughlin in 1983 except for private wells and septic tanks. One reason is that federal dollars for public facilities had dried up, so it was necessary to find local funding. Construction of small-capacity water and sewer facilities totaling about $12 million was initiated in 1983. However, business and population grew so fast that their capacity was exceeded before newly built utilities came on line. So for 10 years there was a scramble to plan, design, and build almost continuous expansions to the water and sewer facilities. Finally, by 1993, over $85 million had been spent to provide the necessary water and sewer facilities.

One of these projects was the construction of the Laughlin Water Reclamation Facility, with an initial budget of $35 million and a 35-month completion schedule. Key participants were owner Clark County Sanitation District (CCSD); general contractor Western Summit Constructors, Inc. (WSCI); the engineer, Black & Veatch; and several key subcontractors. FMI facilitated the partnering process.

It was evident from the start that this was going to be an extraordinary project. First, consider the cultures that the two key players, Clark County Sanitation District and Western Summit, brought to the partnership. Jim Gans, director of the CCSD, feels that the most important contribution his organization made was an unwavering commitment to partnering: "We are 100 percent committed to partnering, as we see no other alternatives for improving the way we do business, and we will not accept any half-hearted efforts."

WSCI's culture, as described by their president, Ed Wambsganss, aligns perfectly with partnering values. As Ed says: "We operate with a goal of avoiding litigation, and to date have never been in court. Our values allowed us to embrace partnering with open arms."

In addition, both parties brought in technically qualified and experienced personnel. Many CCSD team members had previous field construction experience working for contractors. Many WSCI team members were extremely familiar with the operational requirements of the wastewater-treatment plant. This allowed both parties to communicate better, to produce innovative solutions, and to build trust within the team.

Both of these organizations also embrace the continuous-improvement values evident in TQM.

Building Trust through Partnering

The most difficult hurdle in the partnering process is the attitudes the parties bring with them to a new construction project. These attitudes are often the result of years of experiences on previous jobs in which owners and contractors have been "burned." Owners believe that contractors are just looking for ways to "cut corners" to decrease their costs while at the same time searching for flaws or weaknesses in the design which will enable them to propose change orders or present claims—in both cases to increase profit. Contractors believe that owners want them to absorb the costs of any unexpected situations or shortcomings in the design, and purposely leave the specifications ambiguous to force the contractor to do things never envisioned in the bid. What a way to start a project and begin positive relationships! This is why some projects at best confirm old beliefs, and at worst create mortal enemies among project personnel. In particular, the Laughlin project came on the heels of a disastrous Clark County project with a multimillion-dollar judgment and unreimbursable legal fees on both sides that exceeded the judgment.

How can this hurdle be overcome? By using the partnering process to open communication, take risks, and build trust. Trust is the key, and building it does not happen overnight. One must be willing to take risks and talk honestly with all partners. Partnering meetings and the partnering processes established during these meetings provide opportunities to build trust, but someone usually needs to start the ball rolling. A small concession on a minor issue is a good start.

In the Laughlin experience, the contractor began the process with an obvious interest and an organized approach to the project, and early submission of a well-thought-out and workable schedule that was actually followed. In addition, the contractor stated a goal to finish the project under bid with no loss of quality. Owner's representative Jim Gans elaborates on how trust building began on the project:

> Our first test came right at the beginning of construction. The contractor asked to be allowed to simultaneously begin construction at multiple locations. This was not contemplated in the job specifications. Not expecting this request, we gulped. The construction site was an existing, fully operational, wastewater treatment plant that was operating almost at capacity. We took a deep breath and agreed to the contractor's request, even though we knew it would have significant impact on our operations and believed it was not a good idea. We came to realize we were wrong. We found the contractor to be very knowledgeable about the operation of the existing treatment facility, the treatment process, and our operation. He used this knowledge to coordinate his activities with our operations, always giving every consideration to minimizing disruption of our activities.
>
> Several times the contractor helped us ready areas for construction work. We were understaffed due to difficulties in recruiting qualified people. If we did not have the equipment to expedite our work, or if we could not get a facility vacated as quickly as we anticipated, the contractor would provide the equipment and operators to help us get it done!
>
> About mid-way through construction, some of our treatment equipment malfunctioned and flooded out a large concrete basin the contractor had recently poured. When the water drained off, tons of mud and debris remained. Because of the configuration of the basin, heavy equipment could not access the mud; cleanup would have to proceed with manual labor. This would cost the contractor loss of critical time and

thousands of dollars in labor. We saw a major claim coming which would be difficult to defend. The claim never materialized. The contractor adjusted his schedule to accommodate the extra work. In addition, CCSD staff assisted in the cleanup effort. The contractor only charged us his cleanup costs which we thought were very reasonable.

On another occasion, our sister agency, the Big Bend Water District which serves Laughlin's potable water needs, had a water main break in the main Laughlin thoroughfare, Casino Drive. The Water District's equipment necessary to excavate the roadway was not immediately available. Meanwhile, Casino Drive was becoming impassable to traffic. The contractor sent a backhoe and operator to the broken water main, dug up the pipe, and assisted in its temporary repair. And when we asked how much it would cost the Water District, we were told, "No charge."

Many other examples confirmed that trust levels were continually building during the Laughlin project. This trust building pays off most when significant problems arise. From the owner's standpoint, the trust became so high in Laughlin that when the contractor said a change order was necessary, it was hardly even questioned. The contractor prepared legitimate change orders with reasonable costs. Many significant change orders were for decreases in the contract amount, which resulted in a final cost only slightly above the original contract amount.

These experiences combined to create a strong bond of trust and mutual respect between CCSD and WSCI.

Winning through Mutual Objectives and Accountability

Tied closely to this concept is the common characteristic of win/win relationships with shared objectives. In implementing TQM, a company shares its objectives openly with all its employees, and employees begin to buy into those objectives as their own and to share responsibility for the company's success. During partnering, employees come to understand the importance of developing common team objectives derived from each project stakeholder. These objectives are often written down in a project charter, to which all stakeholders commit by signing their names. These overriding project objectives facilitate win/win relationships because all stakeholders are resolving issues in a manner that is likeliest to achieve project objectives.

Here is an example of how we used partnering to resolve problems: Early on, a sand lens was encountered under the large concrete tanks. This lens made it very difficult to place, as required, select fill on a 4:1 slope under the structures. The contractor was allowed to eliminate the fill and instead placed a 3-in concrete mat.

When a company or partnering team is developing goals and objectives, they need to be quantified. Next, present status must be measured, so that improvement and degree of success can be determined.

At the initial partnering retreat, the Laughlin participants were asked to identify goals. The stakeholders enthusiastically participated in choosing goals and setting performance objectives for each goal. Seven kinds of goals were chosen:

1. Safety
2. Empowerment
3. Quality
4. Budget
5. Schedule
6. Innovation
7. Fun

See Fig. 6.1 for more detail.

Safety

The cost of workers' compensation and pressures from several outside forces has caused contractors to pay a lot of attention to safety. WSCI has for the past several years made safety its number-one priority. According to Ed Wambsganss:

> A lot of people will tell you they have also, but you need to take a real close look as to whether it is just talk or whether there is visible action.

Goal	Measurement	Objective	Actual Results
1. Safety (300,000 man-hours)	OSHA recordable incident rate Lost-time incident	6.0 0	5.45 0
2. Quality	Number of deficiency reports Problem solving Major punch list Submittal rejection	Less than 23 100% 30 Less than 10%	5 100% 0 7%
3. Budget	Changes to contract price	1.5%	0.2%
4. Schedule	On-time completion	35 months	23 months

Figure 6.1 Partnering goals, Clark County, Nevada, wastewater-treatment plant project. Aggressive goals were set—and exceeded.

My idea as to the key to anyone's safety program is to look at the top and see if the person at the top has totally dedicated him- or herself, or have they tried to delegate that activity. Does that person hold people accountable for their attitudes on safety? To me, safety primarily is an attitude. Either you have it, or you don't. Once you have it, things sort of feed on themselves and before you know it, the attitude starts flowing down around the work force and you have a great safety program.

The Laughlin project set a goal of 6.0 for the OSHA recordable incident rate and a lost-time incident rate of 0. This was a very ambitious goal, given the fact that the industry average incident rate is 13.

By the time the two-day initial partnering workshop was over, the stakeholders had gone a long way to assuring that there would be a safe project. As Ed says:

> Everyone went away with a goal in their head or on paper in the notes they took. I don't want any of you to think that a safety program is that easy. Safety is a very complex, technical issue. You need a professional, trained staff to advise as to all the necessary parts of a good program—fall protection, confined space, and MSDS technical sheets, to just name a few. You can have all of that stuff available, but if the attitudes are not right, your safety program will fail. The bottom line is we did it.

Project results were a 5.45 OSHA recordable incident rate and no lost-time accidents with almost 300,000 labor hours worked by the entire team. The project received two Nevada AGC Safe-Site Awards. A special thanks goes to Bob Gass, the on-site safety director, and to Boyd Dunham, general superintendent, who really had the right attitude on this project.

Empowerment

In partnering, decision making is pushed to the lowest level to achieve the goals of the stakeholders. Time is money, and low-level decision making expedites the completion of the project, which benefits everyone's bottom line.

Empowerment was one of the goals committed to by the Laughlin team members, and was successfully implemented through the use of an issue resolution matrix, shown in Fig. 6.2. Across the top are the organizations involved. The vertical axis shows each party's personnel at each of six management levels, here ranging from first-level supervisors at the top to top executives at the bottom (thus unresolved issues move down the chart).

As Ed Wambsganss explains:

> This is an area I think most all the credit needs to go to our facilitator, Lou Bainbridge of FMI. He brought to our group this idea that most decisions need to be made at the lowest level possible in our organiza-

Clark County Sanitation District		Western Summit/TIC		Black & Veatch	Helix Electric	Hi-Grade Concrete	Western Technology
Inspection: Robin Core J.R. Holcombe Harold Suitor Dave Mehrhoff	Operation: Mike Yonke Rich Hansen	Field: Gene Duran Rick Lyons Clive Waring Pat Rodriguez	Engineers: Bob Sauguinetti Richard Hatchell		Foreman:	Sherry Huntzinker Joe Scheible	
Robin Core J.R. Holcombe Harold Suitor Dave Mehrhoff		Tom Paul		Daryl Meineke Jill Reilly Gary Parton – Electric Rob Jordan – HVAC	Eric Blom	Joe Scheible	Tim Burkhard
Bernie Davis		Boyd Dunham	Mike Gardner	Jill Reilly	Eric Blom	Phil Allen	Tim Burkhard
Jerry Grover (702) 298-1841 County Offices: Punda Pai Hazel Dewey		Dick Kirkpatrick		Boyd Hanzon	Ernest Jeffries	Phil Allen	Paul Bowen
Kent Olson		Mike Graeve		Dana Reel	Dave Gandzik	Terry Thorpe Bob Shafer	Jim McNutt
Jim Gans		Ed Wambsganss	Rod McKenzie	Dan Linstedt	Gary Shekhter	John Hove	Jim McNutt

Figure 6.2 Issue resolution escalation ladder, as used at Clark County, Nevada, wastewater-treatment plant project. The top row shows first-level management of each party, and the bottom row, their chief executives.

tion. This is what made it all work. We were shown that if we kept our decision process and our dispute resolution horizontal, and at the lowest level possible, we would get results.

Lou then showed us that if we could not resolve a dispute at a certain level, it would move up vertically in each organization. This sounds pretty simple, however, I know before partnering we would have had parties that would go up diagonally on the page, which causes problems.

Jim Gans and I were never really forced to sit down and tackle an issue because the system worked and all issues were worked out by management levels below us and most were settled at much lower levels.

Quality

At Western Summit, quality is priority 2, after safety. Clark County also gave quality a high priority and brought the people responsible for quality to the partnering meeting. This project was a very complex civil and mechanical design. There were over 200 pieces of equipment to be installed, and assuring that they met the project specs was critical.

The partnering process again opened up the lines of communication to discuss issues and allow those people involved to offer suggestions. For example, Black & Veatch agreed to allow some communication between the parties prior to its rejecting an item of equipment. A goal was set that fewer than 10 percent of these submittals would have a flaw that would require a resubmittal.

As Table 6.1 shows, ambitious quality goals were set—and exceeded.

Budget

How many times have you heard of a job coming in under budget and ahead of schedule? Most times, owners set up a contingency fund or an allowance for changes that amounts to 5 to 10 percent. It quickly became apparent that Clark County wanted the final price paid to be very close to the original contract price.

On a complex project such as this (mechanically complex, and with the requirement to keep the existing plant operational at all times),

TABLE 6.1 Quality Goals and Results

	Goal	Actual
Deficiency reports	<23 (1/month)	5
Problem solving (without litigation)	100%	100%
Major punchlist items	30	0
Submittal rejection	<10%	7%

something can easily be overlooked during the design or planning stage that will require change. Once orders are placed and construction starts, it becomes very difficult to change things without incurring additional costs.

Kent Olson of Clark County stated that he wanted to see less than a 1.5 percent change from the contract price. The stakeholders were again involved to come up with ideas and a plan as to how to accomplish this goal. Many people believe that contractors "low-ball" their bids to win jobs, and then make their profits from change orders. Western Summit does not use this approach even if it believes the plans will require many changes. Needless to say, if this had been the case, this part of the partnering process would have failed miserably. Engineer Black & Veatch had prepared a well-designed set of original drawings.

The plan included receptiveness to cost-reducing changes that would offset scope additions. Everyone agreed that project quality could not be lowered, and that, of course, the engineer and the owner had final say. The other factor that began to come into play at this point was trust building. This cannot be emphasized enough. If the reader takes only one idea from this case study, we hope it is that success depends on building trust. One can work hard at it for months, and totally destroy it in pricing one change order. Partners do not try to gouge one another, and the Western Summit team always felt that Clark County wanted them to be profitable.

The original contract was for $35,117,300. The net amount of change orders added $304,781. Of that, $230,000 was caused by added scope, so the project came in at $75,000, or 0.2 percent, over budget, well below the team goal of 1.5 percent.

Schedule

The schedule is important to every project. In treatment plant construction, long durations have been common. On this project, the contract set duration at 35 months, with several milestone dates along the way. There must be good communication lines between the fabricator, general contractor, engineer, and owner or there will be significant delays. Partnering opens up those communication lines, so the whole process flows much more rapidly. It became possible to have all the materials and equipment available to the construction crews as they were needed.

CPM scheduling is a very powerful tool if it is utilized properly. It provided a very detailed road map as to what the needs were as the project progressed. It helped achieve completion in 23 months, early by 12 months or nearly 33 percent, and substantial completion occurred two months prior to that. The teamwork attitude was the reason.

Problem solving and innovation

Partnering relies heavily on proactive problem solving to address issues before they become crises. Employees learn quickly that objectives can be met only by using such an approach.

Without structured problem solving, the success rate for achieving project objectives is greatly diminished. The Laughlin project utilized a weekly problem-solving structure to resolve problems before they became issues.

For example, at one point, seven-day concrete breaks began to come in low. A meeting was immediately called which all involved parties attended. Before casting any more concrete, we agreed on what to do: additional tests, changes in mix design at the batch plant, and so on.

With objectives and problem solving comes a need for innovation and risk. Continuous-improvement philosophies rely heavily on innovation to generate creative solutions to opportunities for improvement, and employees soon learn that "if you always do what you've always done, you'll always get what you've always gotten."

Big gains require innovative ideas. Employees who are accustomed to stretching the limits in a TQM program—not to feel overwhelmed by difficult objectives, but to keep pursuing them—will be innovative on a partnering project. They will have less fear of failure because their organization has taught them that successes will be rewarded, that these successes frequently involve risk, and that failures will be used only as learning tools.

This, in turn, encourages them to share risk with their fellow project stakeholders in order to realize more dramatic results in achieving project objectives. Many of the examples in this Laughlin project exemplify shared risk and innovation.

The contractor, because of the trust developed, felt free to question the design, and to offer innovative solutions such as beginning construction in several areas at once. Another example stands out: The basin concrete pours had been designed for a steplike process that Western Summit felt would be unsuccessful given the sandy soil conditions. Structured problem solving did not generate a solution. Finally, Western Summit agreed to assume its share of the risk by absorbing the cost of an additional 300 yd^3 of concrete. Their solution was successfully implemented. The owner was eventually responsible for $30,000 in additional reinforcing steel, but felt comfortable with the risk they had also assumed in choosing this solution.

Communication is essential to any successful business entity, and a driving force behind successful implementation of TQM and partnering. In TQM, employees are given the tools of successful communication, including facilitator training, meeting management, and overall leadership skills. They are taught that communication is vital to understanding objectives, attacking opportunities for improvement, and empowering employees. Partnering encourages a free flow of information that allows for informed decisions, made in a timely manner, which in turn reduces project time and wasted resources. The tools these TQM employees bring with them to a partnering project are invaluable in generating this type of communication. The stakeholders in the Laughlin case attribute much of their success to the free flow of information and to the honest relationships that developed among the participants. As Jim Gans stated, "Everyone returned phone calls, and you knew you could communicate with anyone if the need arose."

Many partnering teams wrestle with setting a project goal of having fun. Some think you cannot tell people to have fun, and others will argue that "This is my job/life and I am going to enjoy myself." Having fun and recognizing individual and team successes is important in the partnering process. Everyone wants to be part of a winning team. Sometimes we take ourselves much too seriously. Human nature is to be critical, and at times that is all we are. We need to set a goal to rise above this tendency. A quality job and having fun are not mutually exclusive goals. To the contrary.

Lessons Learned

A company that embraces the values of continuous improvement will find that employees are better able to adapt to the partnering process and often play more of a leadership role within the partnering process.

Managers have an understanding of what is expected of them throughout the partnering process, because they have already been taught many of the skills necessary for a successful partnership, including well-developed skills in innovation, risk sharing, communication, decision making, team building, issue resolution, and critical success factor measurement. These skills are essential to successful implementation of partnering.

The Laughlin Water Reclamation Facility project was successful because of the values and attitudes brought to the table by the major stakeholders. Their ability to build trust, take risks, and communicate effectively led to win/win relationships that benefited all stakeholders.

As Jim Gans states:

> It should be obvious that while this construction project involved the typical construction challenges of most any other project, it was atypical in the successes with which the challenges were resolved through partnering.
>
> In fact, it was so successful that at the conclusion of construction we decided to formally recognize and thank the contractor for an exemplary

project. Only one or two contractors have been brought before the owner's Board of Trustees for such recognition in the past two decades. An engraved plaque was created for this occasion, citing the project's early completion, the safety awards, and the cost savings initiated by the contractor. At the public meeting, Board members expressed their appreciation to Mr. Wambsganss. The concluding remark from the Chairman of the Board (just as Mr. Wambsganss was leaving the podium) best captured the spirit of this appreciation. "Ed, do you do roads?"

Characteristics of Successful Partnering

1. Team-oriented culture
2. Atmosphere of trust and respect
3. Empowerment—decision making and issue resolution at the lowest possible level
4. Win/win relationship based on jointly developing project objectives
5. Measurement of critical success factors
6. Problem-solving focus by partners
7. Innovation
8. Openness to risk taking and risk sharing
9. Open and honest communication

Chapter 7

The Partnering Process

A Project Management Strategy to Improve Quality and Field Productivity

Lou Bainbridge

Lou Bainbridge describes two keys to partnering success: (1) transforming project parties into a team whose members pull for each other; and (2) getting project teammates to attack the inevitable disagreements and problems, aggressively and systematically, and not each other.

This chapter relates how one project partnering team developed a "Project Covenant of Good Faith and Fair Dealing," which listed among the parties' mutual duties: "Each party will assist in the other's performance; each will avoid hindering the other's performance."

Of course, even partnering teammates do not always agree. When they cannot, they need to agree that they disagree, and move the issue up to the next project management level. At the start, the team develops a chart that defines who is responsible in each organization, at each level (project, division, headquarters, and so on). Senior management requires escalating all unresolved issues up this management ladder. When partnering does not work well, it is sometimes a sign that this issue escalation process is not being used properly.

Partnering As a Management Tool

Overview

Partnering is about money and how the team chooses to manage its resources. All parties or "stakeholders" must understand the financial parameters of the project and the impact that their interrelationships will have on those financial issues. By focusing on preventing obstacles, the team members will be less likely to have to choose between maintaining profitability/budget and cooperation. Figure 7.1 shows how a partnering team working together can find high-value opportunities during the design and construction process. Prevention of problems leads to cost-effective design and construction.

Partnering is not a legal term, but a management strategy that can be implemented once you have understanding and are prepared for hard work. The driving philosophy behind the success of partnering is the concept of good faith and fair dealing. Simply said, this philosophy requires that we will be helpful to each other and make business decisions in keeping with the team's mission and goals. To highlight and reinforce this way of doing business, some people are adding narrative in the project's special provisions. An example of this is shown in Fig. 7.2.

This provision emphasizes the two parts of any contract: the legal requirements and the working relationships. The partnering process puts a new emphasis on the working relationships and the values of trust, respect, and honesty.

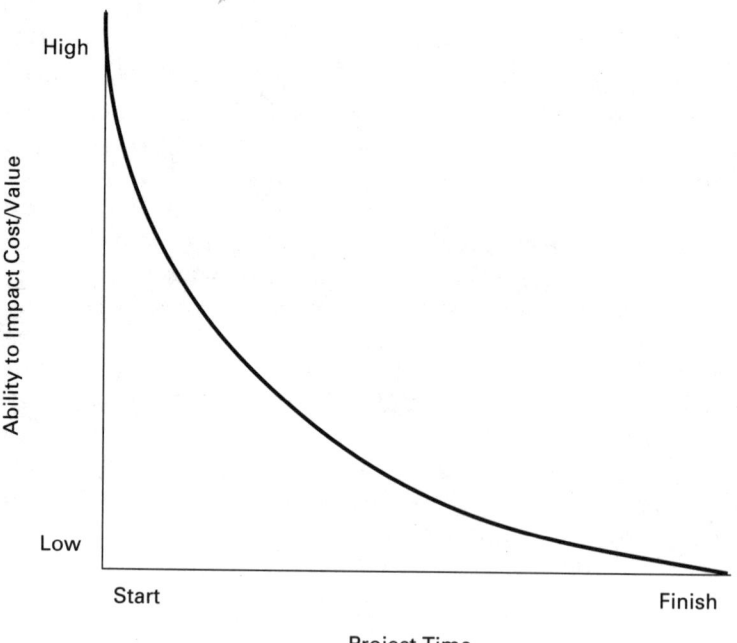

Figure 7.1 The benefit curve. One's potential to impact a project is greater, the earlier in the job you act.

> This contract imposes an obligation of good faith and fair dealing in its performance and enforcement.
> The contractor and the Owner, with a positive commitment to honesty and integrity, agree to the following mutual duties:
>
> A. Each will function within the laws and statutes applicable to their duties and responsibilities.
> B. Each will assist in the other's performance.
> C. Each will avoid hindering the other's performance.
> D. Each will proceed to fulfill its obligation diligently.
> E. Each will cooperate in the common endeavor of the contract.

Figure 7.2 Covenant of good faith and fair dealing.

The process and tools

Partnering is a structured process developed to produce a successful job for all involved. It focuses on the working relationships so that the overall project can be designed and constructed in the most cost-effective manner. There are four major phases: preparation, the partnering workshop, field implementation, and project close-out.

Preparation phase. Senior management must be fully aware of what will be implemented and the resources required for them to be truly bought into the process. If partnering is to fail, it will begin to unravel at this step. Issues that need to be explored during this phase include overall project goals, potential risks, and roles and responsibilities of the various organizations involved. The plan forward must include agreement and understanding in those areas as well as agreement as to who is going to sponsor the partnering process at the executive level and the job-site level.

Preproject partnering workshop. Most design and construction requires a large expenditure of time and money in a concentrated period of time. Each project should be viewed as a significant enterprise, requiring an appropriate amount of business planning and coordination. This first partnering workshop accomplishes just that, typically during a two- to three-day period of time.

This session focuses on how the stakeholders will work together and manage change throughout the life of the project. In addition to the standard information about the project, people, and extensive project-specific problem solving, some fundamental tools are developed during this meeting to help the partnering teams throughout the job. Partnering consists of a broad spectrum of activities, but the minimum includes the project charter, a team report card, an issue resolution process, and problem solving.

1. The *project charter* is probably one of the most visible tools during the project. The mission statement and objectives contained in the charter are the project team's commitment to how they will work together on the project. The stated objectives will be used later as a benchmark to measure how the team is performing against initial expectations. The charter should be posted in a prominent location to remind all participants of their partnering commitments. Figure 7.3 shows an example of this tool.

2. The second tool of partnering is the *team report card*. It is used to measure the performance of the team and to encourage mutual accountability to team goals. An evaluation survey should be filled out by partnering team members monthly. The feedback from this survey will be used to identify opportunities for improvement and to document successes of your team. Specific rewards and recognition can be established to reinforce team performance. An example is shown in Fig. 7.4.

3. The *issue resolution process* is a definition of how the team will resolve issues when teammates disagree on solutions to problems or when a personal conflict arises. In a partnering process, the tone or manner in which the team deals with issues and with each other is important. Documentation should be comprehensive, yet it needs to be brief and should reflect joint discussions. The tone should be courteous and free of threats or attacks. In all cases, issues or conflicts require immediate and face-to-face discussion and action planning.

When teammates cannot agree on an issue, they need to agree that they disagree, and move the issue up to the next management level on the project. The team will develop a chart that defines the principal communication points for each organization as well as the levels in the issue resolution process. Senior management will require escalation of all unresolved issues. If partnering does not work well, it is

---

We pledge to work together to construct a high-quality, environmentally sound, and aesthetically pleasing Water Reclamation Facility. The following objectives will be achieved

- A Safe Job
- Meet Budget
- Quality Work
- Innovation
- Meet Schedule
- Clear Communications
- Maintain Positive Relationships

Figure 7.3 Project charter for the Laughlin, Nevada, Reclamation Facility.

Team Goal: A Safe Job	Goal	Annual
1. Incident rate (OSHA)	< 6.0	5.45
2. Lost-time accidents	0	0
3. OSHA citations	0	0
4. Material loss/damage	0	0

Team Goal: Meet Budget	Goal	Annual
1. Change orders	< 1.5%	.05%
2. Profits		as bid or better

Team Goal: Quality Work	Goal	Annual
1. Number of deficiency reports	< 1/mo.	5 (total)
2. Major punchlist items	30	0

Team Goal: Innovation	Goal	Annual
1. Submittal of deduct charge orders	75% accept	100% 15 submitted; worth $337K

Team Goal: Meet Schedule	Goal	Annual
1. Major equipment target dates	100%	100%
2. Major equipment delivery dates	100%	100%
3. Meet CPM milestones	100%	100%
a. All phases ahead of schedule		

35 months	contract
26 months	team schedule
21 months	actual completion

Team Goal: Clear Communications	Goal	Annual
1. Rejected submittals	<10%	7%
2. Unresolved claims	0%	1
a. Mixers/clarifiers issuer		

Maintain Positive Relationships	Goal	Annual
1. Issues escalated above level four	0	0
2. Conflict intervention	< 4/mo.	0
3. Qualified personnel turnover	< 2%	0

Figure 7.4 Typical team report card.

probably a sign that the issue escalation process is not being utilized properly.

4. The *problem-solving phase* of the initial workshop defines specific obstacles to a successful project. Also summarized are action plans for solving the key obstacles. To ensure success, teammates must follow up on the action plans created. The list of "rocks in the road" generated at the workshop should be used to continue proactive problem solving throughout the life of the project. A partial list of "rocks in the road" from a project is shown in Fig. 7.5.

Field implementation. For the process to work, the project members must put the tools into daily action. All workers should be oriented to the team's specific mission and goals. Opportunities for improvement need to be searched out and implemented using a total team approach.

- Unknown conditions behind asbestos
- Insufficient hoist facility and coordination
- Negotiate/execute MODs in a timely manner
- Non-partner involvement which hurts the process
- Noise complaints from tenants
- Tracking signage order and installation
- Changed requirements by agencies
- Default by subcontractors
- Turnover of key people
- Labor strikes
- Quality control of finishes
- Timely approval of shop drawings and submittals
- Understanding of authority levels
- Prompt punchlist identification and completion acceptance
- Government-supplied materials
- Discrepancies between contracts 3 and 5
- Safety issues
- Inadequate funds for tenant improvements
- Lack of design requirements for floors 3 through 5
- No authority for field changes
- Cost overruns by change orders
- Delay in progress payments
- Lack of understanding of how the agency works—the approval process
- Lack of tenant focus
- Schedule delays—extensions due to delays in defining scope
- Lead paint
- Change order turnaround and cash flow
- Building electrical problems

Figure 7.5 A list of "rocks in the road"—things that the partnering team identified as possible problems on their job.

Value-engineering incentives help drive innovation at rates greater than without partnering.

At predetermined milestones, additional partnering planning meetings should be conducted. Progress to date should be reviewed, new subcontractors and vendors should be included, and additional problem solving should be conducted. These meetings typically last from one-half day to two days. Many people tie project celebrations into these meetings as a way to recognize and reward the measurable results accomplished to date.

Partnering close-out. As the project draws to a close, the stakeholders should reconvene for a partnering process close-out meeting. During this meeting the team should review the people side of the project as well as performance against the team goals established. Specific measurement criteria should be reviewed. Positive aspects should be noted, and opportunities for improvement should be outlined. The team's focus should be to identify process improvements as well as specific design issues which could have led to improved performance. Each organization must take this input and build action plans for improvement. This is a great source of practical ideas to support an organization's quality efforts.

What Do Participants Think of the Process?

We define formal partnering as a process involving preparation, a workshop, and planned follow-up activities throughout the duration of the project. Workshop activities involve the development of partnering tools including a project charter, team report card, issue escalation process, structured problem solving, and behavioral profile.

In August 1993, FMI sent survey questionnaires to personnel involved in 200 projects where FMI facilitated a formal partnering process. Surveys were received from 114 projects—an excellent response of 57 percent.

Eighty-five percent of the respondents were public works projects, and 15 percent were private. These numbers are not surprising when one considers the following:

1. Most of the work available in the industry in recent years has been in the public sector.
2. Frustrations revolving around the construction process have accumulated over a long period of time on hard-bid public projects.
3. The public sector has had a visible partnering champion in Charles Cowan, through his work with the Corps of Engineers in the Portland District and the Arizona Department of Transportation.

The majority of respondents (68) were highway projects. The remaining 46 were building, industrial, and other miscellaneous projects.

Eighty-seven percent of the projects responding fell into either the $0–$10 million contract size or the $10–$50 million contract size. This is a fairly accurate representation of the size of most contracts in the industry. The total dollar value of the projects was $4.3 billion.

There was an even distribution of projects in terms of percent complete, with the range being 5 to 100 percent complete. Sixteen projects were 90 to 100 percent complete.

Respondents were asked to provide information in the following areas:

1. Scheduling
2. Budget control
3. Safety
4. Quality
5. Working relationships
6. Issue (conflict) resolution

The survey was directed toward project-level teams and personnel. Information on the above areas was collected in three formats: quantitative project data, subjective rating scales, and open-ended comments. The subjective rating asked respondents to compare the partnered project with past, nonpartnered projects on a 1–5 scale, with 1 being "much worse" and 5 being "much better."

Overall, the results of the survey were consistent regardless of factors such as project size, type of work, percent complete, owner perspective, construction management perspective, designer perspective, or contractor perspective. This indicates that the partnering process is applicable to any project situation. However, partnering is no panacea—a serious level of commitment must be given to the process by senior leadership of each of the participating stakeholders.

Scheduling

Seventy percent of the projects were reportedly on or ahead of schedule. The 1–5 subjective survey average, when compared to past projects, was 3.4; see Fig. 7.6. Weather had been a factor among those scoring in the "worse" or "much worse" categories, although several respondents said that partnering helped get the project back on schedule.

Positive comments indicated that partnering had increased "joint scheduling efforts" among the parties, creating better buy-in from all

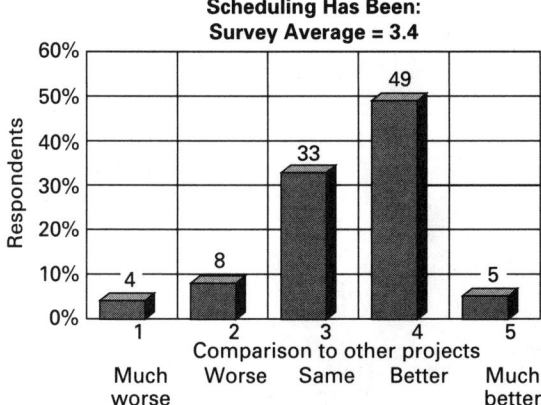

Figure 7.6 The impact of partnering on project schedules: 54 percent reported that it helped, 12 percent that it did not.

concerned in a true bottom–up approach. Industrial project contract language changed to accommodate joint scheduling. The intent of the change was not to hold one party responsible for the creation and creativity involved in scheduling, but to make this a joint responsibility so that all would be involved. This led to more openness and innovative thinking on the many ways the project could be completed successfully.

Team scheduling is an important process in increasing productivity on the job site. Partnering is slowly converting the CPM schedule from a project weapon that is used to build cases for claims to the powerful planning tool it should be. Open communication is essential if any scheduling process is to work. This means that all parties should be viewed as equals in the schedule discussions. No one group should feel that the schedule is being forced on them. Our results showed that the industry needs to increase subcontractor involvement in the scheduling process.

Budget control

Eighty-two percent of the projects reported being on or below budget. A total savings of $96.5 million was reported on the $4.3 billion of work, or 2.2 percent; see Fig. 7.7.

Inspection costs were reduced as a result of increased trust. Reduction in time spent on defensive case building can show up on the bottom

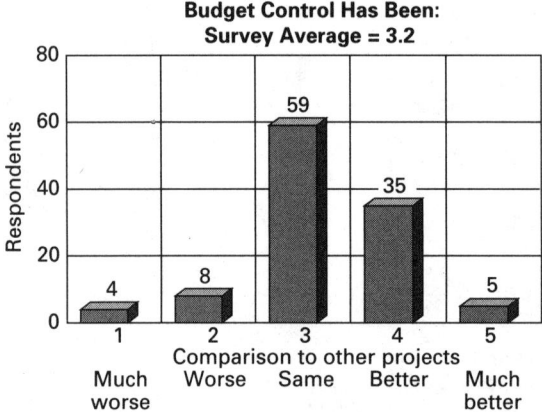

Figure 7.7 The impact of partnering on budget control: 40 percent said that it helped, 12 percent that it hurt.

line of a project budget, as personnel devote their time to more productive activities.

Partnering is all about managing money. Every project has a finite amount of funding. Our experience indicates that the partnering process will increase the rate at which cost issues are identified and resolved. However, less than 40 percent of the projects surveyed actually improved in the budget area.

Negative comments suggested that industry paradigms such as "partnering cannot work in a hard-bid environment" and "change is costly" are still factors on some projects. The fact that 85 percent of the projects in this survey were public works shows that partnering does work, regardless of procurement methods. The negative perception of change orders can be overcome when cost savings are produced.

Some projects had budget control that was adversely affected by ineffective project administration. This is characterized by problems such as not enough people to do the required work and/or levels of approval authority not being clearly defined or coordinated.

To further increase the effectiveness of partnering, senior management must encourage the project leaders to do the right thing when it comes to work performed, risk assumed, and results achieved. Projects must be managed by fact rather than emotion. Too many projects continue to underperform because of one or two unresolved financial decisions.

Safety

There was very little debate over the issue of safety. Safety was a primary focus for project teams even before partnering was introduced.

Respondents stated that partnering safety performance on these projects ranged from the best they had seen to the "only bright spot" on a very difficult job (see Fig. 7.8). Safety is a total team effort.

It was frequently mentioned that a greater empathy was developed on the part of all stakeholders for third-party (public, business/landowner) concerns regarding safety issues. One of the best results recorded was an Occupational Safety and Health Administration recordable rate of 2.3 on 1 million labor hours. This openness to work together aggressively to achieve measurable results in safety gives the partnering process great potential. Achieving this goal lays the foundation for future job teams to become comfortable in their commitment to other goals, such as scheduling and budget.

Quality

There were no reports of negative impacts on quality due to partnering. This is significant and should ease the fears of those who believe that partnering is a "giveaway" program to contractors in which enforcement of specifications is eased, items are overlooked, and quality is sacrificed. When using partnering, the contract and specifications are always enforceable.

Partnering helps decipher gray areas of the contract and specifications, while assisting project teams in mutually agreeing to the requirements that will produce the desired results.

Sixty-five percent of those who provided open-ended comments

Safety Performance Has Been:
Survey Average = 3.6

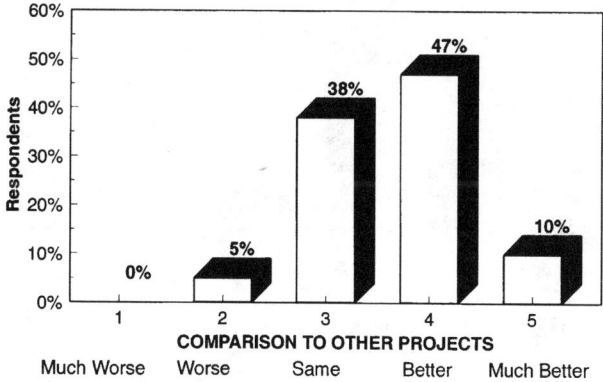

Figure 7.8 The effect of partnering on safety: 57 percent said that it helped, 5 percent that it hurt.

cited specific quality improvements on their projects or had positive things to say about the quality of the work. Thirty-five percent said that quality was the same as on other projects, or indicated that it was too early in the project to have an opinion.

Working relationships

Eighty-two percent of respondents reported improvements in attitudes, trust, and relationships due to partnering, or experienced positive relationships on the project (see Fig. 7.9). Only 8 percent experienced negative relationships. When compared to past projects, the 1–5 subjective survey average for teamwork was 3.6 (see Fig. 7.10). Fighting over traditional territorial issues and hidden agendas was reduced.

Negative comments noted that an environment of trust had been difficult to achieve on some projects. Corrective actions by senior management were lacking. The softer issues of communication, trust, respect, and teamwork are excellent leading indicators for the project's ultimate schedule and budget. More industry executives need to monitor these softer factors if they truly want to know the pulse of a project.

Surprisingly, many partnering teams wrestle with having a goal of fun. Some think you cannot tell people to have fun, and others will argue that "This is my job/life and I am going to enjoy myself." Having fun and recognizing individual and team successes is important in the partnering process. Everyone wants to be part of a winning team. Sometimes we take ourselves much too seriously. Human nature is to

**Trust and Respect Have Been:
Survey Average = 3.6**

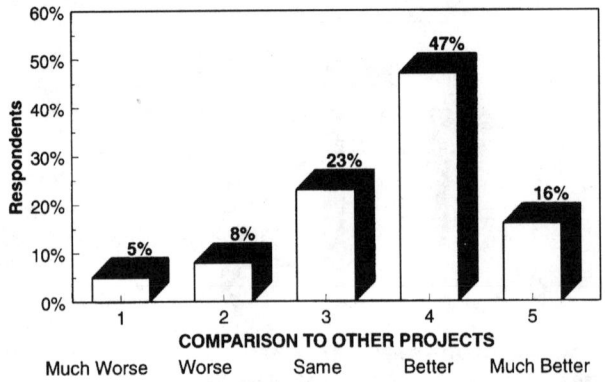

Figure 7.9 The impact of partnering on trust and respect among teammates: 63 percent said that it helped, 13 percent that it hurt.

Teamwork Has Been:
Survey Average = 3.6

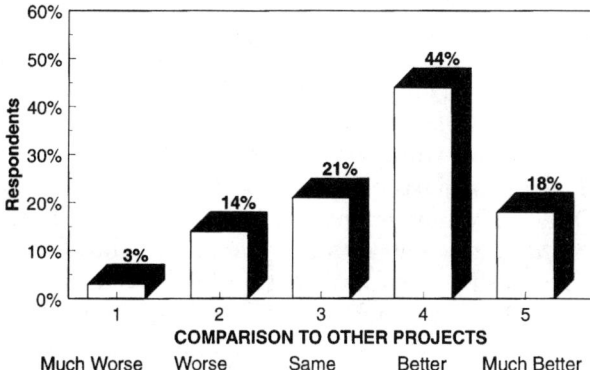

Figure 7.10 The effect of partnering on teamwork: 62 percent said that it helped, 17 percent that it hurt.

be critical, and at times that is all we are. We need to set a goal to rise above this tendency.

In order to build a team, we need to actively look for and document successes. This sets a tone of opportunity which even on a bad day will help carry the team to a successful conclusion. Many projects establish team logos, hand out special items (hats, shirts, etc.), and/or conduct barbecues for the workers to celebrate. On one project more than 40 individual recognition certificates were awarded to team members during the life of the project.

Issue (conflict) resolution

Seventy percent of respondents reported improvements or positive experiences regarding issue resolution. Compared to past projects, the 1–5 scale average for the time required to resolve issues was 3.5. Positive comments suggested that issues revolving around change order, submittal, and request for improvement processes are still difficult, but are not being ignored. These processes are frequently the focus of problem-solving efforts at the initial partnering workshop. While negotiations can be difficult at times, partnering keeps the parties focused on business and the mutual objectives of the team by sorting out these issues.

Other comments reinforced the notion that a large majority of problems should be resolved at the job-site level rather than at the upper-management level. By defining specific roles and responsibilities of each team member on the project, issues are resolved faster and, more times than not, at the project level. Project teams are noticing the difference with its use of the partnering process. Since early 1992 the Texas Department of Transportation has reduced the number of claims from an average of 28 claims or formal disputes per year, with an average $4.4 million paid, to only one claim worth $200,000 and no formal disputes since the inception of partnering. Another example is the Corps of Engineers' claims, which have dropped from 1103 in 1986 to 532 in 1993 as a result of partnering. These results have caused the White House's Office of Management and Budget and its Office of Federal Procurement to endorse alternative dispute resolution techniques such as partnering. The jobs have not gotten any easier; it is the use of formal partnering tools and enlightened leadership that is helping produce these outcomes.

Where problems with escalation still existed, a common theme expressed by respondents was that the resolution process had been discussed at the initial workshop but never implemented on the project. To overcome this, we suggest testing the escalation process in the first month following the kick-off workshop. Early in the process, the executive management team for the project should look for issues that are particularly problematic or unresolved. Management needs to act as a coach or facilitator to see that the team clearly defines the problem and develops specific action plans. The team problem-solving process and the resulting outcomes need to be highlighted for the entire project team. This will showcase the benefits of proactive behavior and send a very positive message to all participants.

Overall, it appears that project teams escalate technical issues in an expedient manner, but fail to escalate the people issues in the same manner. The industry seems to put more emphasis on finding fault than on resolving issues. Hence, trust and relationship issues continue to be a factor on some jobs. The construction industry must recognize that this is a people business and that existing behaviors are hard to change. It will require a concentrated management effort to reinforce the vision of the future and ensure that we are living the defined team values. A portion of each partnering meeting should be dedicated to discussing the people side of the business. Specific skill sessions, such as how to conduct meetings, listening, effective communication, and the like, should be part of the ongoing process.

Completed projects

Typically in the construction industry, once a job is complete, the project team disbands and goes its separate ways. This made it difficult to collect data on completed jobs. We were able to gather the following from approximately 10 completed projects:

1. Project size ranged from $2 to $300 million, with most projects in the $2–$10 million range.
2. On average, these projects were completed 5 percent below budget and 18 percent ahead of contract schedule.
3. The responses to the teamwork/communication trust issues scored higher than those projects we surveyed that were still in process. There was a keen sense of accomplishment.
4. All project teams said that they would use the partnering process on future projects.

Obviously, in these cases the partnering process worked; however, all the teams realized that there was room for improvement in certain areas, including schedule, safety, budget control, and/or teamwork. It is not uncommon for project teams to score themselves low in an area such as safety after one or more instances where someone was almost hurt, or in issue resolution where an issue was not handled properly. This "rating down" is not unusual among high-performing teams. The knowledge of opportunities for improvement is one way to spur superior performances on future jobs.

Would you partner again?

When asked the acid-test question, "Would you partner again?" an overwhelming 105 of the 114 projects (92 percent) answered "yes." Of the seven projects responding "no," the following thoughts were offered on why partnering did not work as well as hoped:

1. Weather was a factor.
2. Trust/respect was poor.
3. There was a lack of take-charge people in critical decision-making roles.
4. Participants had been "forced" to partner.

The most common reason was lack of leadership. One project responding "no" was just 15 percent complete. It is difficult to understand how partnering can be fairly evaluated so early in the process unless this team had negative experiences during the preparation and workshop steps. We need to reemphasize that partnering is *not* a two-day workshop, but an ongoing process throughout the duration of the project.

The two projects that responded "unsure" to partnering offered the following comments:

1. All entities must be committed to make it work.
2. Some contracts contain too many defense mechanisms to provide a true partnering atmosphere.

Critical Success Factors

Partnering within the construction industry has been a tremendous success. There is brilliance in its simplicity, yet we must commit as an industry to improve our project execution. Partnering does not eliminate the problems, but it does provide a system for teams to identify and resolve issues at the lowest possible level. We cannot tolerate business as usual if we are to be successful. Partnering is not just team building. Our industry must improve all project processes to have cost-effective money management. The following is a list of critical success factors.

1. Partnering must be a strategic initiative. The organizations that have experienced the most success with the process have made it a long-term strategy for improvement that does not isolate successes on just one job. Total quality management mechanisms are in place, so good ideas developed on partnered jobs can be easily transferred to other projects. Good ideas are not lost when the project ends.

2. Senior management must be committed and actively involved. Top-level executive officers provide the leadership vision for their organizations by participating in the process. Participation in the preparatory and workshop stages is critical. Reinforcement of the issue escalation process and other follow-up activities during the course of the project is necessary. Where partnering has not worked well, frequently it is because senior management provided nothing more than "lip service" to the process.

3. Reasonable expectations of what can be accomplished need to be established in the beginning. Partnering does not eliminate all the problems on a project, nor does it create a perfect world. It is an evolutionary process that helps define how problems will be dealt with. If there is some question as to the commitment level of one of the key stakeholders prior to initiating an agreement, this concern should be addressed in the preparation phase. Thereafter, team members should focus on what they have control over if they hope to avoid disappointing results.

4. The process should begin in the design phase, before construction start-up. The ability to affect behaviors and business practices is

reduced if partnering is introduced later in the project. Few projects to date have been formally partnered during the design phase. Respondents indicated that this would help establish a positive tone for business relationships before bringing the contractor on board.

5. Time and resources must be committed. Like any process, partnering does not work as well when it is short-circuited. This requires both a personnel and a financial commitment. Managers should view the process as a business decision that is an investment versus a cost. An investment should have clear returns associated with it. Senior managers should understand the tangible results that occur. If they are still unconvinced and are unwilling to commit the proper resources, they should not be partnering.

6. Job-site personnel need to be included in the process from the beginning. It is important that foremen, field engineers, and inspectors be brought into the process on the same level as senior management. At least 80 percent of the issues originate on the job site. Attitudes need to be adjusted at the front line if the process is to work. One shortfall of the partnered projects was not involving subcontractors, architects, engineers, and all their respective subconsultants. Success is difficult to achieve if these players do not participate in the initial workshop(s).

7. Follow-up and reinforcement are required to sustain the process. *Partnering is not a two-day workshop.* The process is an attempt to evolve behaviors and attitudes that have developed over many years. It is unrealistic to think that this change can occur within a 48-hour time period. It is important to plan reinforcement, so that partnering is a part of day-to-day job-site activities and vocabulary. Weekly/monthly meetings with key project leaders to discuss the process and to do problem solving are a must. Longer projects (6 months plus) should plan additional team workshops to reemphasize partnering tools, to conduct more problem solving, and to bring new players on board.

8. Measurement and monitoring of team goals are essential. Project leaders should review these goals as a part of their follow-up partnering meetings. This provides the opportunity to identify problem areas early and to correct them before they begin to erode relationships. The responses on our surveys indicate that this industry is still struggling with the concept of measurement. Key quantitative questions in the areas of safety, quality, and budget control were frequently left unanswered.

The key obstacles to measuring team performance are that it takes time and the attitude that "we all know how we are doing." It does take time, but not as much as people think. The data are all there; they just need to be combined into one useful report. With this information the team can manage by fact. More important, all participants in the project can be involved in understanding which goals are being achieved and which have room for improvement. Measurement not only becomes easy, it allows for exceptional performance to occur. (See also Chap. 6.)

An example of exceptional performance is one project that monitored team goals on a weekly basis. The project was for Rockwell and was completed by DPR Construction and team. It was a design/build fast-track renovation of clean-room fabrication facilities. Approximately $39 million worth of work had to be completed in five months: a challenging task, to say the least. Throughout most of this project there were two to three shifts per day working six to seven days per week. The total team effort exceeded 150,000 labor hours. The project was far from perfect, and the team was continually looking for ways to improve. It was this commitment at the project level that produced exceptional results. Figure 7.11 highlights the final outcome.

9. Adhere strictly to the issue resolution process. If partnering is not working well, often it is because the escalation process is not being utilized properly. While empowerment for decision making is an important part of partnering, project leaders need to understand that if they cannot come to closure on a job-site issue, it is acceptable to escalate that issue off-site. Sometimes project leaders will hold onto an issue to the point where personal relationships begin to erode. Senior management must reinforce and participate in the resolution process.

10. All parties must be accountable for their responsibility on the project. An atmosphere needs to be created in which stakeholders can admit their mistakes and not feel they will be "hammered" by the other parties. All parties need to realize that it is in the project's best interest and the interest of all concerned to help a team member through times of trouble.

11. Give partnering plenty of visibility. The more people understand and feel a part of the process on the job site, the more they will contribute to the success of the project. There are still many situations where the crews hit the job site and believe it will be business as usual. This indicates a big opportunity in the industry to further impact safety, quality, schedules, and budgets with total involvement of all job-site personnel.

Partnering Trends

Today, partnering is becoming a common practice on many projects. If partnering is to survive beyond the next several years, however, the

Operations Goal

Item	Goal	Actual
Fabrication yield loss	0	0
Probe yield loss	0	0
Missed moves (rework)	0	0

Quality Goal

Item	Goal	Fab 4	Fab 5	Fab 1	Central plant	Bulk chem
Punch list items with operation impact	0	0	0	0	0	0
Design deficiencies resulting in cost impact	0	0	0	0	0	0

Quality Goal

Item	Goal	Actual average	Reward recognition program
OFIs implement per employee	1	1/2	Project team shirts, lunch barbecues

Schedule Goal

Item	Schedule date	Actual date
Fab 4 tool hook-up complete 4/28/94	4/28/94	
Fab 5 ready for tool hook-up 6/10/94	6/10/94	
Fab 1 ready for tool hook-up 6/24/94	6/23/94	
Tool hook-up complete	8/5/94	8/4/94

Contractor Coordination Goal

Rework in field, item description	5/1	5/15	5/29	6/5	7/10
Hanger rods before fireproofing Bulk Chemistry	1				
Ceiling in Fab 1 set too low		2			
Gel leaks in Fab 5 ceiling grid			1		
Cooling tower sump basin leak				3	
Mechanical piping cross supply/return					2

Figure 7.11 The very aggressive goals (and impressive performance) on a Rockwell clean-room job by DPR Construction.

Safety Goal

Goal	Actual	
0	11	Incident
0	5	Recordable accident
0	0	Lost-time accident
0	5	Other (non-injury accident)

*Note: Approximately 150,000 labour hours expended

Information Flow, Close-Out, and Building Department Relationships Goal

Item	Goal	Actual
Submittal review	3 days	4 days
RFI review	4 hours	48 hours
Charge order pricing	5 days	9 days
Close out complete	9/5/94	9/2/94
Permit red tags	0	0
Lost time due to reinspections	0	0

Payment Performance (Days)

Payment number	Rockwell to DPR	DPR to subs	Subs to suppliers
1	13	N/A	N/A
2	10	15	20
3	9	14	20
4	23	4	18
5	13 partial	4 partial	15
6	13 partial	4 partial	15
7	22	5	15
8	Pending	Pending	Pending

Figure 7.11 (*Continued*) The very aggressive goals (and impressive performance) on a Rockwell clean-room job by DPR Construction.

industry will have to get serious about investing the time and money to implement ongoing processes properly. The benefit will come from project control systems that will continue to be streamlined as more and more systemic problems of information flow are corrected. Involvement of subcontractors and suppliers as equal partners on the project team will improve overall job performance.

Construction insurers, bolstered by partnering's ability to reduce claims and litigation, will encourage the adaptation of job-site partnering to minimize errors and liability for omissions. The surety industry will encourage the use of partnering as a way to ensure that an ongoing management process is in place to keep the project on track.

As partnering becomes more of a strategic initiative within owner organizations, we will see a more common-sense approach to the implementation of the intent of the plans and specifications. Particularly in the public sector, we will see officials take responsibility for making valid judgments on rules and regulations instead of deferring to others. Although this accountability entails some personal risk, leaders can use the systems of partnering to show measurable benefits to support the actions they take.

In the future, the industry will continue to find ways to involve people in identifying and implementing solutions to stretch design and construction dollars. We are only just beginning to tap the potential of the industry's ability to be innovative and efficient throughout the project team.

Part 2

Other Project Partnering Applications and Implications

Chapter

8

Case Studies in Strategic Alliances

Anthony F. Costonis

In what is called strategic partnering, industrial clients are making two changes: (1) cutting their number of contractors by up to 90 percent; and (2) with each of those remaining, using total quality management or a similar quality program, with the goal of progressive improvement. Among clients of management consultant Anthony Costonis of Corporate Development Services (Lynnfield, Massachusetts) are contractors working with industry-leading Intel, Ford, and DuPont.

Costonis reports: "When you're inside the cultures of these contractors you instantly know something is different. Most of their staffs and employees are highly motivated—turned on."

The Construction Industry Institute (Austin, Texas) has begun benchmarking the savings from multiproject strategic alliances, and recently reported that 196 projects built using long-term partnering saved an average of 15 percent of total installed cost.

America's elite corporations—represented in this study by DuPont, Intel, and the Ford Motor Company—are changing the way they do business with the construction industry. They are not only demanding more for less, they are getting it. This objective is achieved by their commitment to quality and continuous improvement, both in their internal programs and externally in the strategic alliances they develop with their suppliers and vendors.

Each corporation has developed a slightly different approach in the selection of "suppliers of choice." Each program is described in detail, indicating what contractors must do in order to receive certification and qualify as a "supplier of choice."

The benefits to the contractor—represented in this study by a group of ABC Delaware Chapter contractors (Wilmington, Delaware), an Oregon electrical contractor, and Sylvan Industrial Piping (Pontiac, Michigan)— are a significant improvement in internal productivity and enhanced market position.

The implication to the construction industry is the need to recognize that buying philosophies of corporate America have changed, and the change will significantly redefine the competitive landscape of the construction industry.

Introduction

The following item appeared in *USA Today* during the third week of August 1994:

> Intel plans to build a $1.3 billion semiconductor manufacturing plant in Hillsboro, Oregon, by 1998 and spent $700 million to expand one in Aloha, Oregon. The Hillsboro plant would eventually employ 350 people and another 350 would be added at Aloha. Intel is already building plants in New Mexico and Arizona that are scheduled to begin operating in 1995 and 1997.

It was only several weeks prior to that article's appearance that Ford Motor Company announced the highest quarterly profits ever reported by any automobile manufacturer.

Meanwhile, in Wilmington, Delaware, at the headquarters of DuPont, after five consecutive losing quarters, the company had bounced back with record-setting profits in its last two quarters.

Few would argue that these three companies are among America's elite corporations. Ford and DuPont have long and storied histories. Intel, on the other hand, is a relative newcomer to the *Fortune* 500. Each, however, is now at the "top of its game," operating at productivity, efficiency, and profit levels higher than ever before.

Why? What do these three corporations (and others experiencing similar success) have in common?

The Secrets of Success

Each of the three corporations has embraced, without reservation, all of the concepts that comprise what has come to be known as "the quality movement." Inherent in their commitment to quality is the drive for continuous improvement, an effort that a company can achieve only in concert with its suppliers.

To borrow from the introduction to Intel's Supplier Continuous Quality Improvement (SCQI) manual:

> Quality is co-owned (by both customer and supplier). The true costs associated with poor quality are now better understood and it has become apparent that improving quality is in the best interests of both participants. Adversarial relationships have been replaced by cooperative effort to achieve ongoing improvement in all facets of supplier quality from defect-free products, error-free services, and highly reliable equipment with maximized output to responsive delivery and competitive pricing.

DuPont, Ford, and Intel all agree that one of the first, and most important, steps toward achieving the kinds of relationships we have just described is to be highly selective in the choice of suppliers. From their existing pool of suppliers, only those who can meet specific criteria will qualify. If you are one of those selected, you are presented with a substantial opportunity. However, if you are not, you face the prospect of not even having a chance to compete for a substantial piece of corporate America's construction business.

Industry has adopted the term *convergence* to describe the weeding-out process. Convergence often takes on massive proportions, as shown in the data from *The Wall Street Journal* that are reproduced in Table 8.1.

How has this affected the construction industry? Companies like Ford, DuPont, and Intel now buy construction services in terms of value-added criteria rather than just on price. By adopting the "don't buy on price tag alone" concept of the acknowledged founder of the quality movement, W. Edwards Deming, they seek to do business with companies that have the same commitment to quality and customer service as they do. Whereas many construction buyers perceive the services our industry provides as a commodity, forward-thinking

TABLE 8.1 Strategic Partnering: Effect on Number of Suppliers

	Number of suppliers		
	Previous	Current	Percent change
Xerox	5,000	500	−90
Motorola	10,000	3,000	−70
Digital Equipment	9,000	3,000	−67
GM	10,000	5,500	−45
Ford	1,800	1,000	−44
Texas Instruments	22,000	14,000	−36
Allied Signal	7,500	6,000	−20

SOURCE: *Wall Street Journal.*

companies, like those noted, have recognized the fallacy of such a belief.

In this chapter we shall take an in-depth look, through three case studies, at how specific members of the construction community have responded to the changing, more demanding circumstances at Intel, DuPont, and Ford. I and my firm, Corporate Development Services (CDS), have been extremely fortunate to have the opportunity to work with a number of progressive contractors who have adapted to the new buying philosophies of these three leading corporations.

Each of the three manufacturers takes a slightly different approach to building strategic alliances in their attempt to achieve continuous improvement with their suppliers and vendors. They represent three means to the same end. Despite the differences, there are common threads running through all three. It is hoped that these case studies will provide guidance and direction for other construction companies that have the intelligence and are willing to make the commitment to participate in the future as it is being defined by these companies.

Delaware ABC Takes the Lead at DuPont

The DuPont Corporation spends about $3 billion annually on construction. According to Russ Eckerson, DuPont's manager of corporate contracting, "Awarding contracts to the lowest bidder wasn't getting us where we wanted to go" in terms of completing high-quality jobs on schedule.

As a result, DuPont set out to reduce the number of contractors it does business with by 85 to 90 percent, from some 800 to 1000 to approximately 70 to 90, focusing on those contractors that are committed to total quality management and have such programs up and running in their companies. DuPont estimated the savings to be $15 to $25 million annually in the Wilmington area alone, and $200 to $300 million a year worldwide.

In designing its convergence program, DuPont contacted the Construction Industry Institute (CII), the industry's "think tank," to develop the quality criteria that would be used to select the contractors. DuPont established five criteria:

1. Capabilities—organizational chart; capabilities and personnel; market covered; workload mix; company strengths and weaknesses; in-house work versus subcontracted work

2. Quality—how quality, productivity, and cost are measured; how schedules are monitored; type, depth, and coverage of quality training; hiring qualifications; professional/business involvements

3. Partnering/competitiveness—how competitiveness is maintained; women and minority efforts; safety record
4. Financial—Dun and Bradstreeet rating; number of liens filed against; sales and profits; expectations as a converged contractor
5. Safety/health/environmental administration—safety program and training; substance abuse testing; accident investigation; awards and incentives; safety and quality matrix

Next, DuPont generated a list of all the contractors it had done business with over a period of years. It did a preliminary screening and then invited the remaining contractors to respond to a detailed questionnaire and interviewing process.

For example, in the electrical trades, DuPont began with 118 contractors in the Delaware area. Preliminary screening reduced the number invited to participate in the qualification process to 53.

Using responses to a rather detailed questionnaire as its basis, the process ultimately reduced the final number to 11, who were designated "alliance contractors," and that now are responsible for all DuPont's electrical work. This represented better than a 90 percent reduction in the number of electrical contractors that will now be doing DuPont's Delaware-area work.

Responding to DuPont's moves (and similar actions by other large Delaware-based corporations), the Delaware Chapter of Associated Builders and Contractors called in my firm, Corporate Development Services, to help its members become part of DuPont's new, slimmed-down team. CEOs from 35 different ABC member firms invited CDS to conduct a two-day workshop on total quality management (TQM). After this workshop, the contractors empowered the chapter's board of directors to appoint a quality committee to work with CDS to develop a program that ABC members could present to DuPont as evidence of their commitment to quality.

CDS and the chapter established a four-level "Excellence in Construction" certification program:

Q-1. Certification designed for a company that wants to learn what TQM is all about at the CEO and senior management level, including some introductory training. This level is also designed to explain what steps need to be taken if a company is to implement a TQM program.

Q-2. Certification designed for a company that has made the total commitment to TQM and wants to design and install a TQM program to involve its employees in the process of continuous improvement with the assistance of CDS's trainers and facilitators.

Q-3. Certification designed for companies that wish to establish strategic alliances or partnership relationships with customers, vendors, or subcontractors.

Q-4. Certification designed for CEOs of companies that have already been certified at the Q-2 level of achievement. It provides a forum for exchanging ideas on lessons learned, especially with reference to "best practices" and benchmarking techniques.

Ten contractors, including general contractors, heavy/highway, mechanical, electrical, and other subspecialty firms, participated in the Q-1 and Q-2 programs. Nine of the 10 were selected as "alliance contractors" in the DuPont program.

The Delaware ABC Chapter continues to offer the TQM program, reinforcing quality as a valuable tool for competing effectively in a market that not only seeks, but demands, excellence.

SCQI Rules at Intel

Unlike DuPont's more shotgun approach, Intel zeros in and is much more actively involved in contractor assessment activities. Though it is only a small looseleaf binder, Intel's Supplier Continuous Quality Improvement manual is the bible for any supplier or contractor that wants to do work with the company.

A "prerequisite" phase plus three formal phases encompassing a total of 10 steps are involved. If a contractor is successful in moving through the program, the company becomes a "supplier of choice." As we shall see, those that remain standing at the end truly deserve to be there.

The prerequisite phase stimulates a company to determine whether it is ready to approach the SCQI process. Before plunging ahead, prospective suppliers are called upon to understand what Intel wants, who it wants it from, and to come to an agreement with Intel on how to get there—a true multiproject partnership alliance.

Phase 1

Having passed initial screening (a process similar to DuPont's convergence milestone), a supplier moves on to phase 1, "Planning and Preparation." Nicknamed "Getting Your Act Together," phase 1 is a three-step process that involves creating a clear and concise road map for the implementation of SCQI.

Step 1 is setting expectations—defining where you want to be. To quote from the SCQI manual, "Failure to achieve the required results from a good supplier is most commonly attributable to a poor job on the part of the customer in stating clear expectations." Intel is saying, "If you, Mr. Contractor, are struggling, we take it as our challenge, too."

Step 2 is current situation analysis, or in other words, establishing a baseline of where you are now. Are you heading in the same direction as Intel? Are your goals and objectives consistent? Do you have the resources to meet Intel's needs? In addition, benchmarks are established and your existing quality system is audited.

The final step (step 3) of phase 1 is to develop a plan—a clear strategy to enact SCQI. Using the information gathered previously, an SCQI road map is established featuring scope, objective, schedule, estimated resources, and priorities. It is then distributed to all the involved parties, and becomes a living document, updated and revised regularly.

Phase 2

Titled "Supplier Qualification," phase 2 has a number of goals: to reduce costs, delays, product development time associated with poor quality and variability in the product or service; and to build confidence in, as well as build, a system that puts ownership of the continuous improvement process in the hands of the supplier.

Step 4, "Alignment," is the process of obtaining complete understanding and agreement on Intel's requirements. Through a series of meetings that harken back to the expectations of step 1, agreement is reached—either by the supplier accepting the conditions defined by Intel or by Intel modifying the conditions so they are acceptable to the contractor. Alignment must establish the boundaries of what is acceptable and what is rejectable.

Step 5 collects data and analyzes it in those areas where Intel's requirements are not being met, and step 6 brings the contractor into complete conformance with what Intel requires, resulting in consistent, reliable adherence to the documented requirement.

Phase 3

The third phase of SCQI, "Supplier Management Improvement," goes to the heart of continuous improvement. The contractor is now the primary quality driver, with Intel merely providing validation and guidance. A contractor must be actively operating in phase 3 in order to be an ongoing supplier to Intel.

Step 7, "Self-Assessment," is a contractor-performed annual in which Baldrige Award–like standards are used. That self-audit is validated by a site audit performed by Intel (step 8).

Once validation has taken place, a jointly prepared improvement plan, with specific objectives, is drawn up (step 9), followed up by periodic reviews two to four times per year (step 10).

If all this sounds tough, it is. But those who survive the process reap great rewards. Consider, for example, an electrical contractor and Intel vendor in Hillsboro, Oregon.

The contractor was on the TQM bandwagon early, in the late 1980s. The firm began its relationship with Intel by doing a few small contracts. The contractor is now building on its TQM foundation by working with CDS to link its quality program to a strategic plan. This involves a conscious effort to make sure quality pays. This is accomplished primarily through benchmarking both corporate and job-site performance, including bid-to-win ratios, job-turnover ratios, and return-on-overhead investment ratios, among others. The benchmarking of these items has led to the identification of improvement opportunities. Quality improvement teams have been established, resulting in a win/win situation for both the contractor and Intel. The contractor's business plan objectives are based on zero defect performance criteria. That is, the contractor has targeted providing continually improved services to Intel.

As a supplier to Intel, the contractor has to be on its toes and at peak performance every minute of every day. Intel expects it—and the contractor delivers it—in measurable quantities day in and day out.

Ford: Quality Is Job 1

Unlike Intel, whose SCQI program addresses all suppliers through the same evaluation process, Ford Motor Company has designed a program specific to the construction industry: the Construction Rating Standard for Contractors. Using this standard, Ford's quality experts visit the offices of contractors that want to do business with Ford, in order to conduct an exhaustive and detailed two-day, on-site audit, labeled Ford's "Construction Supplier Quality Evaluation" (SQE).

The audit has four components: quality, technical, delivery, and commercial. Each of the four components is broken down into subcategories, each of which is assigned a maximum point value. Each contractor's points are tabulated, so the contractor can be ranked.

Though the process is quite detailed, it is worth looking at what Ford evaluates.

Quality

A contractor's commitment to quality is examined carefully. Planning programs, review processes, written procedures, training and education, team organization, statistical analyses of performance, monitoring and improvement systems, and customer liaison are all examined and rated.

Rounding out the quality category, internal and external indicators of the quality operating system are looked at, as are the commitment of the contractor's leadership and human resources and its innovative approaches to procedures and documentation.

Technical

Design is explored in detail—staffing, specific technical expertise, project management, the use of value engineering, project coordination, regulatory compliance, system design/installation, the design and review process, scheduling, budgeting, problem solving, construction planning, dispute resolution, safety policies, computer-aided design capabilities, and serviceability.

Next, the SQE looks at development—improvement initiatives, specification standards and updating procedures, service operating and repair procedures, programs to keep pace with new technologies, continuous improvement practices in staffing and support, and post-project evaluation.

Lastly, project execution/construction is examined—the process flow for problem resolution and changes; how project owners are dealt with; how O&M manuals are developed and how as-built conditions are determined; safety procedures; labor relations; scheduling; review, tracking, and coordination of shop drawings, materials, and equipment; subcontractor coordination; architectural/engineering liaison; and punchlist processes.

Delivery

For the delivery category, once again it's a soup-to-nuts evaluation. How does the contractor ensure that customers' specifications are met? How does senior management communicate with employees? Are you fully computerized? How are project milestones tracked against the schedule? How are critical schedule components expedited? What system is used to resolve customer problems? What about records maintenance and tracking?

Commercial

In the final SQE category, commercial aspects, the contractor's cost competitiveness, management and financial strength, and other business issues are examined. Ford wants to know how price competitiveness is ensured and what cost-saving steps are being taken. It looks at corporate mission statements, operating policies, business plans, organizational charts, and financial statements.

Lastly, the auto manufacturer explores integrity, minority and legal issues, and responsiveness to problems.

In establishing its alliance with a contractor, Ford identifies a number of contractors with whom it wants to work. It then charges these contractors to engage in a corrective action, continuous improvement program based on a mutually agreed-upon review and rating system. The process begins with the contractor rating itself in the four categories identified above: quality, technical, delivery, and commercial. Ford then does its own evaluation and identifies opportunities for improvement. Contractors then submit to Ford a "Corrective Action Plan." Following the same format as the SQE "Construction Rating Report," it details those areas where improvement is needed, what action is being/will be taken, who the team leader will be for that action, and a completion date for each action.

For instance, Sylvan Industrial Piping, a large industrial specialty contractor in Pontiac, Michigan, which has been doing business with Ford for a number of years, has just completed the audit. Sylvan invited CDS to assist it in implementing the corrective action program required by the firm's Q-1 supplier status. Upon review of Ford's requirements, CDS worked with senior Sylvan management to develop a series of alignment steps to ensure effective compliance. These steps include the development of team leadership (including steering committee selection, policy and logistics, selection of teams and tasks, senior management and team training, and facilitation of team-building efforts); vendor certification (including internal assessment, major supplier assessment, and formulation of procurement procedures and policies); and reengineering of core business processes such as estimating, purchasing, project management, job delivery, etc.

The entire effort is designed to involve all of Sylvan's employees in the quality process—particularly challenging in light of the traditional union affiliations common to the automobile industry. Yet Sylvan's leadership, staff, and field employees have embraced the challenge enthusiastically and look forward to becoming "best in class" through the Ford program.

What Does It All Mean?

We have seen the lengths to which three of America's most successful corporations go in order to select their suppliers. The process does not start and stop with them, however. All the companies listed in Table 8.1, and hundreds of others throughout American industry, have been doing the same thing.

As contractors' customers become more sophisticated, they are spending their construction money more intelligently. Their expectations for architectural/engineering/construction suppliers have moved to a higher level. Whether it is called TQM, multiproject partnering, strategic alliances, or vendor certification, it is clear that owners have created a new way of doing business.

Consider the experiences of some of the contractors we have examined in this chapter. Ralph Degli Obizzi, Sr., of Degli Obizzi and Sons, Wilmington, Delaware, mechanical contractors, under whose ABC Chapter presidency the ABC/DuPont program was launched, believes that his experience is an example of what doing business in the 1990s and beyond is all about. As a DuPont Alliance Contractor, his firm's TQM efforts gained his company formal recognition from DuPont by saving the chemical company 5 percent during 1993. "And our profits were up, also," he adds. "TQM is really about saving the customer money as well as making money for yourself. Many contractors are fooling themselves, thinking that they don't need to make this commitment to quality."

And what about Intel and the Oregon electrical contractor? The president of this highly motivated and very successful firm says, "We save the cost and uncertainty of otherwise having to bid for a lot of work to customers who don't share the same commitment to quality we do. We are secure with the relationship we have with Intel, and have found that the quality lessons learned in working with them are paying dividends with our other clients." To top it off, the contractor has been named one of the 10 best companies to work for in the state by the *Oregon Business* magazine, not to mention receiving honorable mention as one of the state's 10 fastest-growing companies.

Sylvan Industrial Piping is proud of its status as a long-time service provider to Ford, a relationship that dates back through two generations of Sylvan management. "Today," says President Mike Morrissey, "we find ourselves excited by the prospect of not only formalizing the alliance, but ensuring that it will continue on over the long term as we become a supplier of choice with an enhanced market position, not only with Ford, but with other clients. Q-1 is more than an advertising slogan you see on TV. It's a vital link in our business and it is a credit to the relationship we share with the Ford Motor Company."

Each of our profiled contractors has made a substantial commitment to quality. Doing that has not been easy, but when you are inside the culture of these companies you know instantly that something is different. You see that they have vision, they are intelligent, flexible, and innovative, and most of all, their staffs and employees are highly motivated to do their best. Each has benefited from making continuous improvement a way of life within their own organization. Stories of increased profits, high employee morale, work as fun, pride of workmanship, and the like are the rule.

Conclusion

As more of the story unfolds, the evidence supporting the benefits of the quality revolution to the construction industry become increasingly compelling. The Construction Industry Institute has begun benchmarking the savings from multiproject strategic alliances. Its Strategic Alliance Task Force reported recently that 196 projects using long-term partnering saved an average of 15 percent of total installed cost. It cited the example of Belcan Engineering Group, of Cincinnati, which reported that it saved a total of $140 million over the last six years through six strategic partnering relationships (that is, compared to the alternative of performing the work on a project-by-project basis with multiple suppliers).

The marketing implications of these results and the many others that might also be cited are terribly significant. Buyers want more for less—and they will get it. However, they will only be willing to pay for it with their suppliers of choice, such as the contractors discussed in this chapter. These contractors have already been at work at continuous improvement for three to four years. They are way ahead of many other contractors who have insisted that TQM is just another buzzword, or those who claim, "we've been doing this for years." How does someone new to these processes compete with those who have already gotten a huge head start?

Already, 20 percent of America's contractors do 80 percent of America's $400 billion worth of work. Given the impetus of the quality revolution, we believe this chasm between the haves and have-nots will widen dramatically in the future. The message is clear: For contractors to win in the competitive environment of tomorrow, they will have to adopt a new vision. Only by making their customers more competitive will they survive, grow, and prosper.

Some contractors, like those we have talked about, have already figured out where the world is heading, and are taking the steps necessary to make their customers low-cost manufacturers and high-profit, high-market-share producers. These contractors have come to understand the benefits of being strategic business allies offering quality rather than alternative suppliers selling price. The lesson they have learned is: "It's not only our market share that counts, it's our customers, also."

Surely, their stories demonstrate clearly the significance of the CII's famous quote in 1990: "Companies must institute TQM or become non-competitive in the national and international construction and engineering markets within the next five to ten years."

Ultimately, contractors like the Oregon electrical firm, Sylvan, and the Delaware ABC firm make their customers the focal point of how they manage their businesses through TQM. They have discovered, as it is hoped many more contractors will, that it is not saying "We're the best," that matters, but instead, "We make our customers the best."

For More Information...

To get a good feel of where corporate America is going with quality programs of this kind, we recommend that you subscribe to *Quality Progress,* (800) 248-1946, Milwaukee, WI, or *Quality Digest,* (800) 527-8875, Red Bluff, CA.

Should you wish to pursue business opportunities in this marketplace, you should contact the purchasing departments of the large American or multinational industrial firms to determine what you must do to be certified.

As the case studies in this chapter have shown, each firm tends to have its own particular or even unique approach to the process. This uniqueness can often make the difference as to whether or not your firm will become a supplier of choice.

Chapter 9

Your Lawyer as a Project Team Member

V. Frederick Lyon

Fred Lyon, managing partner of the Washington, D.C., law firm Lyon & McManus, recommends to his fellow construction lawyers:

- *Make sure contracts allocate risk equitably.*
- *Attend the preproject partnering retreat, and listen.*
- *Let your client know you will consider that you and they have failed if litigation results.*
- *Tell your client, if they refuse partnering, to get a new lawyer.*

Partnering—the key component in the brave new world of construction—is a concept which the industry has wisely demanded in response to the apparently endless dispute and litigation of the 1980s. Many in the industry think that, if it is done properly, partnering will eliminate dispute and its associated bottom feeders, the attorneys. Unfortunately, modern construction is much too complex to eliminate dispute completely. Human nature and the pursuit of profit preclude that possibility. Partnering instead is a technique to *manage dispute*. Consequently, it can and should make use of lawyers as part of the process. Partnering can change the way construction utilizes lawyers, and in doing so make both the industry and its counselors more productive and profitable.

Lawyers should be involved in creating and then implementing the partnering relationship. Partnering will work only if contract documents are risk-sharing and not risk-shifting devices. Dispute is controlled by managing risk; *risk must be borne by the party best able to control it*. Lawyers who are properly informed of partnering objectives, and who are involved early in the project partnering process, are the best tools to ensure that contract documents are drafted equitably to promote partnering. Harsh risk-shifting clauses are inconsistent with the team-building approach. If your lawyers recognize this and draft the contract accordingly, a project is more likely to be partnered successfully.

Another essential element of successful partnering is ensuring that the contract incorporates *dispute resolution procedures* which complement the process and discourage litigation. Job-site (low-level) resolution in accordance with carefully delineated and observed time deadlines is critical. Dispute review boards, structured negotiations, and mediation should be required by contract; litigation must be a last resort. Lawyers who promote partnering can assist in structuring these procedures and then ensure their proper implementation. Dispute cannot be avoided; it can only be controlled. Lawyers have the training and ability to create that control. They are a resource that should not be ignored simply in the haste to remedy the effects of otherwise bad lawyering.

Proactively involving lawyers and encouraging their team-building participation does not mean, however, that there are no legal risks inherent in partnering. The concept is sufficiently new and unfamiliar that it inevitably will be the subject of further definition by litigation within the next few years. The seemingly insatiable creativity of plaintiffs coupled with the unpredictability of judges not experienced in construction ensures that decisions will result which undermine the cooperative principles of partnering. The very use of the term "partnering" suggests to some that the resulting relationship is a "partnership," which is a legal fiction fraught with potential new and expansive obligations. Careful contract drafting will be necessary to ensure that partnering does not promote the very problems it is intended to cure. Sensitive and experienced counsel can prepare contracts accordingly.

The Lawyer's Traditional Role and How It Has Contributed to the Industry's Problems

The American social and economic paradigm of the 1980s and 1990s is unfortunately too well known. It is an economy and a society which have increasingly emphasized the shifting of blame and the avoidance of responsibility as its primary governing principle. The consequence has been a cult of victimization: No one is ever responsible for his or her own problems; it is always someone else who "did something" (or failed to do something). The construction industry, not surprisingly, has hardly been immune from this distressing phenomenon. In fact, given the multiple players on any construction project, inevitably there is plenty of opportunity to be a victim and to blame others for the problems on the project. Be it the owner, the lender, the general contractor, the subcontractors, the suppliers, the surety, the designer, any and all are available as part of the blame game. The result of this game is always the same—dispute—which is all too often resolved on America's favorite battleground, the courthouse.

Another complicating characteristic of American business during the last two decades has been an overemphasis on short-term profits, with less concern for long-term issues. Construction companies, in fact, too often focus on project profits without a willingness to absorb the bad contract; too many project managers have come to believe that one bad job will be fatal to their career. It is better to let the courts decide who is at fault than to accept personal responsibility. Emphasizing short-term profits promotes disputes.

Uniquely situated to benefit from this short-sighted approach to business are America's lawyers. The adversary system is predicated upon finding fault; truth is hammered out on the anvil of differences. The adversary system is ideally designed for a fault-based economy, which explains in no small part the dramatic increase in the number of lawyers during the last decade and a half. In an economy which emphasizes victims—who is at fault?—legions of lawyers will thrive in answering the question. Attorneys feed off the refusal of people and companies to accept responsibility, personal and financial, for problems which frequently are of their own making.

The impact of this approach on construction has been well documented and painfully experienced. During the last decade and a half there has been a dramatic increase in the number of attorneys who call themselves "construction lawyers." Some estimate that the number has increased 400 percent during that 15-year period. The opportunity to make mischief has thus increased exponentially.

Finding fault in the construction industry can be particularly profitable for lawyers. Most construction disputes are fact-intensive, frequently requiring "rebuilding" a major project after the fact. Multiple interviews with numerous participants (most of whom would rather be elsewhere) are necessary. When trying to sort out responsibility, lawyers also have the profitable opportunity to review extensive documents. After-the-fact dispute resolution depends on after-the-fact document review. Any major construction project generates thousands of documents. Trying to assign responsibility by reviewing these documents is inevitably a time-intensive and therefore costly proposition.

Construction lawyers make substantial fees when given the opportunity to wallow in documents.

Within this context, however, lawyers are not the cause of the problem, they are more accurately a symptom. Lawyers typically respond to a demand rather than generating a need for their own services. It is only because so many people and so many companies, particularly in the construction industry, refuse to accept responsibility for their own problems, and the need to manage their own personnel and projects, that lawyers have prospered so greatly, much to the chagrin and frustration of the industry. Therefore, controlling lawyers will require controlling the blame game; controlling attorneys will require that contractors proactively assume responsibility for management of projects. Too often, involving lawyers is not a sign of appropriate aggressiveness, but rather represents a failure of management. If partnering is to succeed, companies must confront this unpleasant reality.

An Opportunity for New, More Productive Use of Lawyers

Using partnering for "lawyer-bashing" is simplistic. To assume that partnering is just a technique to eliminate lawyers prevents internal analysis—more of the blame game. In fact, successful team builders recognize that lawyers provide an opportunity to help create a successful partnering program. Recognizing this reality will bring the lawyers into the process and facilitate the proper management of risk and consequently of dispute.

To involve lawyers in the partnering process is to strengthen it. An attorney's training, for example, develops the ability to understand and appreciate both sides of an argument. This ability "to see the gray," to see that both sides may have meritorious arguments, distinguishes a good lawyer. The partnering process, with its emphasis on open communication, trust, and acknowledgment of responsibility, can make good use of this legal training. It is in remarkable contrast to the training of other professionals in the industry—for example, designers, architects, and engineers—who typically believe that there is only one answer, only one solution to a problem.* These profession-

---

*Joseph Jacobs of Jacobs Engineering has said about designers:

> Engineers have the lowest self-esteem of any profession in the world. They're taught that in college. They're taught that their only function in life is to give us the answers in the back of the book. Engineering tends to attract reclusive, introverted people. People who are afraid to face their own humanity, and take refuge in hard numbers: "No one can attack me when I've added up two and two, and it comes to four."

(*The Wall Street Journal,* September 17, 1991). Jacobs' judgment is a bit harsh perhaps, but certainly not descriptive of good candidates for partnering.

als may therefore be resistant to the flexibility inherent in good partnering. Lawyers are nothing if not flexible. Partnering is about devising creative solutions which the right attorneys, perhaps better than others, can enhance.

Lawyers can serve as either advisors or as advocates. Too often the construction industry utilizes attorneys only in the latter role, as the aggressive, no-holds-barred exponent of a particular position. While perhaps emotionally satisfying in the short run, such an approach is seldom ultimately successful, and is always extraordinarily expensive. A good attorney skilled in partnering will be an advisor, not an advocate. To appreciate the differences, consider the matrix shown in Table 9.1, which shows goals set by the parties in two scenarios: litigation (left-hand side) and partnering (right-hand side).

Lawyers as advisors must be involved in creating and then implementing the partnering relationship. Attorneys should be consulted earlier rather than later. Be assured they are a lot cheaper then. The objectives of the team-building approach should be communicated to the attorney with the direction that the contract documents be drafted to reflect an equitable allocation of risk. Lawyers, particularly in-house counsel, may attend the team-building sessions to listen to the parties' mutual objectives. If lawyers are kept on the outside, it is too easy for them to undermine the relationship.

Lawyers work *for* clients. They should be instructed that they have failed if they do not attempt to resolve disputes without litigation. If a lawyer resists the concept of team building, it is time to get a new attorney. If lawyers are not included in the partnering process, however, they may well work to destroy it. Invest them in the process—not only will that keep them on the ranch, it will allow all concerned to make use of their unique skills.

TABLE 9.1 Lawyer's Two Roles—Gladiator and Team Builder

Advocate	Advisor/team builder
Start high/low	Start realistically
Hold fast	Be flexible
Few concessions	Make concessions
Appear unconcerned	Be concerned
Emphasize winning	Emphasize relationships
Make threats	Make deals

Risk Shifting versus Risk Sharing—Making Sure That Lawyers Draft Equitable Contract Language to Enhance the Prospects for Successful Partnering

Risk shifting is a prophylactic technique intended to avoid blame. It is a creature of the 1980s, when owners in particular believed foolishly that shifted risk was avoided risk. Parties to the construction process kid themselves if they believe that if they don't have the risk, they will never pay the price. If risk in a contract is completely shifted to one party *even though* that party is unable to control the risk, the guaranteed result will be dispute, and lawyers at their worst.

Onerous risk-shifting clauses are largely inconsistent with partnering. They include, but are certainly not limited to,

- No-damages-for-delay clauses
- Broad indemnification clauses
- Soils disclaimers
- Any effort to shift design responsibility to the contractor

How can effective team building be accomplished, for example, when the owner can deliberately delay a job, but then not have to pay for the delay? Lawyers should draft contracts to ensure that risk is assumed by the party that can best control the risk. Partnering is doomed to failure if the harsh contracts of the 1980s survive into the 1990s.

A word about design. Incomplete design remains the lawyer's dream—a fast-track, lump-sum project in which the contractor is made to concede the adequacy of a still incomplete design. Contractors build, they do not design. How can they hard-money price something which is not even finished? These contracts from the high-interest early 1980s sold owners the false hope of being able to shift to a contractor responsibilities that the contractor simply had no business assuming. Partnered jobs should acknowledge this reality. Fast-track, lump-sum contracts remain a prescription for dispute. Designers must participate in the partnering process. They must be persuaded to see the "gray" and actively encourage partnering principles, or the process will not work. The design-build contracts which increasingly characterize the industry, and which endeavor to consolidate risk logically, may in fact prove to be more successful vehicles for the equitable allocation of risk than the fast-track, lump-sum contracts of the last decade.

Dispute Resolution Procedures Which Complement Partnering

Dispute will always characterize any construction process. It is as inevitable as human nature and the demand for profit. Partnering provides an opportunity to control, not eliminate, dispute. As part of that control process, it is critical that contracts provide for controlled resolution of disputes. Certain increasingly popular alternative dispute resolution procedures enhance this possibility. Lawyers again should incorporate these into their drafting efforts.

All such procedures have certain characteristics. Low-level resolution of disputes is absolutely critical to successful partnering. Disputes must be resolved as far down the ladder as possible in order to avoid escalation. Low-level resolution obviously requires empowerment of project personnel and a recognition that sometimes such personnel will have to make decisions which may on occasion erode profitability. That is where deemphasizing short-term profits, and emphasizing the successful partnered project, will bear fruit.

Second, any dispute procedure must emphasize timely resolution of disputes, with kick-up procedures to the next level. Deadlines are critical. If a dispute has not been resolved within five to ten days at one level, it should automatically be kicked up to the next level for resolution. The consequence of this contractually mandated approach will frequently be that lower-level, empowered personnel recognize that it is to their advantage to resolve the matter in a timely fashion before suffering the embarrassment of having it kicked up to the next level. Deadlines imposed by contract and disciplined communication resulting from partnering can achieve this result.

On major construction contracts (say, more than $20 million), a partnered project should consider using a dispute review board. These boards, which usually consist of nonlawyer representatives appointed by each party, meet on-site on a periodic basis, to review the project progress. They should be kept informed of the status of construction. Disputes are presented to the board on a timely basis, almost always without the participation of counsel. Decisions made by these dispute review boards, while not legally binding on either party, can be admitted in subsequent litigation. Such admissibility encourages the parties to abide by whatever decision is made by these construction-familiar personnel. In fact, most projects which have utilized dispute review boards report that they have successfully avoided the need for use of the board by their mere presence. The board is itself a disincentive to dispute.

Another dispute resolution technique that complements partnering is structured negotiations. Such negotiations, which move resolution up the corporate ladder in a stepped, timely fashion but mandate such negotiation prior to litigation, have proven to be very successful. Frequently, the involvement of company personnel not intimately involved in the details of the project is an effective way to avoid litigation. Above all else, such negotiation should be creative, flexible, and exhaustive.

If all else fails and the job still ends up in a dispute mode, anything but litigation remains the preferred alternative. Minitrials, mediation, and reliance on private dispute services are far preferable to the nuclear holocaust of litigation. On large construction projects, even arbitration has proven to be unpalatably expensive and time-consuming. It suffers a further disadvantage of providing for no discovery, no rules of evidence, no reasoned decision, no appeal, and it is usually no faster. On a partnered job, the emphasis should always be on finding ways to resolve disputes without relying on litigation or arbitration. Lawyers should cooperate in achieving that goal.

The Legal Side of Partnering—The Great Unknown

Any time new business relationships are formed, it can be anticipated that the legal components of such relationships must inevitably develop. In the context of partnering, it is frequently asked whether a partnering agreement in any way alters the normal contractual relationships among the parties. Is a partnering agreement a contract, and as such, does it create new duties? New duties, of course, can create new unanticipated risks, altering previously agreed-to contractual relationships. If the language is drafted carefully, however, it can be made apparent that partnering does not alter in any way the duties which otherwise exist or are defined by contract among the parties.

It is important that such limiting language be included in both the partnering charter and in the contractual language which established a partnering relationship. Without such language, it is possible that a creative, trouble-making attorney may argue to a sympathetic judge that, by entering into a partnering relationship, the parties in some way want to create new duties or obligations between them. Whenever new duties are created, the opportunity to argue that such duties have been breached increases significantly. Proper language can ensure that this situation does not result.

A component of this question, of course, is whether the partnering charter itself is a contract. This issue can be of particular importance to the individuals who affix their names to the charter. For example, would an injured worker argue that he or she has the right to sue the individuals signing their names to a partnering charter which encompasses the typical partnering requirement that safety is a mutual objective? Bizarre, yes, but how unlikely in a judicial system that allows the recovery of millions from a cup of spilled coffee? Language to avoid this scenario should be included in the contract. However, generally partnering charters themselves are not contractual documents and should not be characterized as such. There is a risk that creative plaintiff attorneys may argue that such charters create new implied duties which give rise to new causes of action.

A corollary of this problem is the possibility that partnering will be construed by naive or inexperienced courts as creating a *partnership type of relationship* among the parties. Partnership is obviously a legal fiction which is independent and distinct from contractual relationships among contracting parties in the construction arena. The difficulty, however, with a partnering arrangement being construed as a "partnership" is that it could then possibly give rise to the argument that fiduciary duties exist among the partners to the relationship. As anyone who has any experience in the labyrinth of American courts realizes, fiduciary duties in turn give rise to the argument that punitive damages are payable for the breach of such a relationship. To avoid this inaccurate interpretation, many suggest that partnering itself should be relabeled as team building. This "new age" management concept has fewer connotations with potential legal ramifications than the word "partnering."

There are in fact other legal issues that over time could evolve from the partnering scenario. For example, designers may have to consider the possibility that by participating in partnering, they involve themselves in the "means and methods" of construction, which they have spent decades avoiding in order to limit their own liability. Again, appropriate contract language can avoid this damaging scenario.

Subcontractual relationships also could get entangled in the reality of partnering. To the extent that subcontractors do not participate in the partnering workshop and the resulting partnering charter, and to the extent that the owner and the general contractor agree that no claims will result during this job, a nonparticipating subcontractor may argue that its rights have been impermissibly compromised. In taking away rights, then, new ones may be created. Conversely, of course, the danger also exists that the subcontractor that participates in a partnering session will assert a direct right of action against the owner. Appropriate contract language must be included as a safeguard.

The owner's participation in partnering may arguably create new duties. As discussed previously, new rights might be asserted by injured workers who hope to avoid workmen's compensation restraints by participating in a suit directly against the owner, who they argue has taken on a safety obligation. It is critical that the parties to a partnering relationship be aware that simply because they have agreed to control risk and dispute through team building or partnering, this does not mean that the parties who are excluded from the process or creative lawyers will not try to undermine the purposes of partnering by arguing that new duties were created as a consequence of the relationship. Careful contract drafting can avoid this scenario. It can be predicted with a fairly high degree of confidence, however, that the nature of the partnering relationship will be defined in no small part by the courts over the next decade. That is the nature of America.

The answers to all of these questions thus may yet be hammered out on the unpleasant anvil of litigation—still better a job, however, that involves lawyers up front to draft the documents, allocate the risk, and structure the process so that the parties, and not judges and juries, determine if partnering is going to be successful.

Partnering will not eliminate lawyers—it requires creative new ways to use them. Lawyers' skills in seeing the gray can make the process work, with an emphasis on equitable risk sharing and innovative dispute resolution. The best partnering contract is a fair contract that good lawyers will recognize and promote. Like the rest of modern America, there are uniquely legal issues associated with a new concept such as partnering and the attendant risk that if things are not done right, the process will not work. If lawyers are consulted early and carefully managed and controlled, they can be an important asset in implementing the partnering process and ensuring that it is an effective tool to control and manage risk in the modern construction process. Ignore lawyers at your peril; control them to your profit.

Chapter

10

Alternative Dispute Resolution as a Partnering Tool

Robert J. MacPherson and Richard H. Steen

Partnering or no, project disputes sometimes arise. In resolving them, parties are increasingly considering less costly and time-consuming approaches than litigation. Authors Robert MacPherson and Richard Steen describe one of them, "baseball arbitration," as used on a hydro project.

The owner and contractor signed a letter of intent to enter into a construction contract. The contractor was authorized to proceed with specific preliminary work, but with the understanding that if financing was not obtained, the owner had no obligation to award the contract, and the contractor would be entitled to be paid up to the maximum amount set forth in the letter of intent.

Unfortunately, the letter of intent did not fully detail how amounts due the contractor would be calculated. When the owner could not get its financing in place and the project was cancelled, it turned out that the owner and the contractor had widely divergent views as to the value of the work performed. Much of it had been mobilization and preparatory work that the owner, if it eventually did go forward, would have to repeat.

Using a baseball arbitration format, the parties selected a single arbitrator and scheduled a time for each side to present its case. They then exchanged their last best offers. When they still could not settle the dispute, they proceeded to arbitration. The contractor was given a *single day to present facts and circumstances justifying its number, as was the owner. Each side then had one-half day to rebut.*

The arbitrator was then given seven days to select either the contractor's number or the owner's number as more reasonable. The arbitrator selected the contractor's number, which was $400,000 higher than the owner's number.

When the dispute arose, the contractor claimed $1.5 million, while the owner claimed it was due $250,000. Baseball arbitration shrank the difference between the two parties' figures by about two-thirds.

What is alternative dispute resolution (ADR), and why is everybody talking about it?

One view of dispute resolution clauses is that they merely encourage contractors to make claims which they would not make if they were forced to bring a lawsuit in court. Still another view holds that ADR is nothing more than an attempt to get you to waive important protections available to you only in a court of law. Yet those who champion ADR say that it is revolutionizing and revitalizing the construction industry.

ADR is somewhat misnamed, in that it is not concerned solely with the resolution of disputes. More important is a recognition that disputes should be prevented if possible, but if they arise, they should be managed effectively and resolved in an expeditious and economical manner, in which the rights and obligations of all those involved are recognized and respected.

The various ADR processes are tools. If they are used properly, they can make your life and your project much easier.

The Evolution of Alternative Dispute Resolution

Many years ago Abraham Lincoln offered the following advice to attorneys, widely quoted in the legal community today:

> Discourage litigation. Persuade your neighbors to compromise whenever you can. Point out to them how the nominal winner is often a real loser in fees, expenses and waste of time.

Unfortunately, Lincoln's advice was largely discarded by parties to construction projects in the 1970s and 1980s, which saw the adoption of litigation tactics that became known as "hardball," "scorched earth," or "take no prisoners." These tactics sought to take full advantage of the adversarial battle that a jury trial is designed to be. Litigants engaged in expensive pretrial "discovery," which more often than not was anything but. Concealment rather than disclosure often became the operative term, and the discovery process generated its own disputes, which at times overshadowed and required more time, energy, and money than the underlying lawsuit. While lawyers can be and were properly blamed for many of the abuses of hardball litigation, the fact is that they did so at the bidding of their clients, who instructed them to prosecute vigorously what may very well have been questionable claims, or to delay the ultimate resolution of claims against which there was no viable defense.

Litigation was also spawned because of the lack of trust among parties to construction contracts. General contractors often found themselves stuck between an unreasonable and unyielding owner who refused to consider valid claims for extras, and subcontractors who attempted to recover for extras from general contractors even though the general contractor had no means of recovering from the owner. And that is only part of the story.

Even without hardball litigators, nonpaying owners, unscrupulous contractors, and claims-conscious subcontractors, construction litigation would still be time consuming and expensive. Consider, for example, the volume of paper generated by a typical construction project—daily reports, shop drawing submittals, meeting minutes, correspondence, plans and specifications, as-builts, and contract documents. Now consider the time and expense of organizing all that data for presentation to a judge and jury, who may or may not have any knowledge or experience of construction. What you are faced with is a costly exercise that may or may not yield the intended results.

For some time, arbitration was thought to be a solution to the problem of construction litigation. Commercial arbitration, which has existed in one form or another for years, involves submitting a dispute to a neutral individual with some knowledge and background in the subject matter of the dispute, who will render final binding and conclusive decisions on the facts and the law from which there is no appeal.

It was thought by many in the construction industry that by eliminating judges and juries who had to be educated in the construction process, and replacing them with knowledgeable individuals from the industry, presentation time would be reduced, leading to equitable decisions quickly and at reduced cost. By eliminating the right of appeal, an ultimate decision could be reached quickly.

In many cases, especially with well-defined disputes, arbitration was effective, but more often than not, and especially in large cases, arbitration proved at least as time consuming and expensive as litigation. Arbitrations spanning 20, 30, or more hearings over a period of 2 years or more were not uncommon.

The problem with arbitration, as with litigation, is that the cost of participating can approach and at times exceed the amount in dispute.

Alternative dispute resolution works because it allows the parties to be creative. Custom-designed ADR procedures often take the best parts of litigation, arbitration, and mediation and apply them in unique ways.

Dispute Management in Contract Documents

Parties who have decided to use partnering as an integral part of their project strategy have taken an important step in assuring an atmosphere of trust and cooperation, open communication, and focusing the project team on achieving mutual goals.

For partnering to be successful, it is necessary that the contract documents reflect the same kinds of commitments, openness, and fairness.

It is important to review the types of contractual issues that frequently give rise to disputes and that make disputes more difficult to resolve.

Three key areas in contracting need to reflect the partnering attitude: completeness and consistency of contract documents; risk-shifting and risk-sharing clauses; and provisions for the prevention, management, and resolution of disputes.

1. *Coordinate the rights, responsibilities, and duties of the parties into an integrated set of contracts.* This is the only way to assure that all project needs are accounted for, assigned as the responsibility of one party, recognized by that party as its responsibility and priced accordingly, and clearly defined, managed, and accomplished.

Attorneys for the parties play a key role at the outset of a project in drafting contract documents to provide an organized, clear set of ground rules. The development of an integrated project-management master plan at the outset of the project is an important ingredient in defining the objectives of the owner, the resources to be utilized to accomplish those objectives, and the interrelationships among those resources.

An attorney drafting a contract between an owner and a designer must have a clear idea of how the designer is going to interact with the various contractors, equipment suppliers, subconsultants, regulators, and of course the owner. When agreements with construction contractors have provisions giving rise to expectations of the contractor that the designer will have certain responsibilities, those terms also must be reflected in the agreement between the owner and the designer.

Similarly, owner commitments to provide support and resources to all the parties should be consistent among the different contracts with the various project participants. To the extent that responsibility for a certain element of project performance is either inconsistent among contracts or, even worse, is the responsibility of no one, and the omission becomes evident only at a crucial stage of the project, the adverse consequences can rapidly become a major dispute.

A project that starts out with a comprehensive project plan, an integrated set of contract documents, and a partnering attitude is a project that is headed toward success, not toward the courthouse.

2. *Share equitably the risks inherent in every construction project.* This is another secret to contract documents that support partnering goals. Fair allocation of risk is always in the eye of the beholder, but unfair risk allocation makes disputes inevitable and frustrates common objectives.

In the 1980s contracting mindset, risk shifting was thought to be the way to prevent disputes. Parties in the better bargaining position shifted risk to other parties in an attempt to eliminate the basis for recovery on a claim or dispute. But the other parties had not agreed to this (or did so only because they thought they had no choice). Litigation was inevitable.

The risks associated with all construction projects—risks of delay, changes, and cost overruns—need to be allocated to parties fairly. *Each risk should be borne by the party that is best able to control it.* For example, risks of differing site conditions can be much more easily investigated by the owner at the outset of the project than left to the contractor's prebid (and uncompensated) investigation. The risk of delays due to reasons beyond the parties' control should not become a burden to that party.

The federal government is still the only owner that as a matter of policy provides broad rights for equitable adjustments for differing site conditions, delays, design deficiencies, and other adverse events on a project. It is no surprise, in light of that philosophy, that the federal government has embraced partnering as a tool to be used on a regular basis. When all is said and done, partnering is a recognition of the willingness of all parties to treat each other fairly.

3. *Provide for the prevention, management, and resolution of disputes.* One of the primary goals of partnering is the elimination of disputes. That goal must be mirrored in the contract documents through comprehensive procedures for preventing, managing, and resolving disputes. Alternative dispute resolution techniques are the tools to

achieve these objectives, and a contract should spell out the procedure for using these techniques to minimize disputes. Many of the more successful ADR techniques are described later in this chapter.

There are still only two basic ways to resolve dispute, by agreement of the parties or through reference to a third party who issues a decision. When the parties cannot agree, binding arbitration is the litigation-avoidance form of choice. Between simple negotiation and binding arbitration there lies a vast area for fashioning techniques and procedures to help the parties achieve agreement. The benefits of agreement are well known: preserving the relationship of the parties, allowing the parties to focus on the project and not the dispute, and minimizing the time and cost to resolve.

Just as the partnering facilitator helps make the partnering charter a reality, a neutral third party whose role is to facilitate resolution of disputes (as opposed to imposing a decision) is a key feature of a forward-looking dispute management process. Contract terms that provide that parties to a project have access to a neutral, on the job site, at the time the potential dispute surfaces, are becoming more prevalent and are universally recognized as crucial to achieving dispute prevention, management, and resolution goals. Dispute review boards, job-site neutrals, project mediators, and other types of neutral facilitators have helped parties negotiate and achieve dispute resolution goals.

Within the hierarchy of project organizations, a key goal is to get the dispute negotiated at the lowest appropriate level within the organization, preferably by the personnel directly involved in the situation. Open communication, supportive upper management, timely access to relevant facts and implications, and a strict deadline for resolution are all necessary components for making the system work.

Comprehensive dispute management provisions include:

- Requirements for notice and invocation of the disputes procedure
- Requirements for documenting the issues that have arisen and the positions of the parties
- The steps in the process and levels of authority
- The framework for the negotiations
- The role of neutrals and the timing of the involvement of neutrals
- The roles of the owner, designer, and contractor in documenting the facts and clarifying the obligations of the parties
- Compensation for the neutral

- The schedule for proceeding through the various dispute resolution steps
- The confidentiality and precedential value of the resolution
- The implementation of the settlement

When the parties to a project are faced with a potential dispute, the availability of a comprehensive procedure for resolving disputes gives those parties and that project important advantages. The nature and consequences of a dispute cannot be predicted in advance, so contractual provisions for addressing the wide variety of disputes which may arise need to keep disputes in their proper perspective, keep the parties on the project focused on key project goals and objectives, and, of course, move the dispute through the process toward resolution.

The absence of dispute resolution provisions frequently results in a dispute becoming more important, at least in the eyes of some project participants, than the project itself. In addition, in the absence of a process there is a tendency to defer problem resolution. The longer you wait, however, the more difficult it is to establish the facts and the more hardened the parties' positions become. The incentives for consensual resolution are often no longer available, and an adversarial battle becomes the only remaining option. It is axiomatic that the longer a dispute is ignored or resolution is deferred, the more costly, time consuming, and difficult it becomes to resolve.

Disputes that are not promptly identified and properly managed can rapidly escalate and may even doom a project or one of its participants. Despite the best intentions of all the parties, not all disputes can be avoided or prevented, and not all disputes can be resolved by the parties without intervention.

Dispute Prevention and Resolution Techniques

Dispute prevention and resolution options can be viewed as forming a continuum, as illustrated in Fig. 10.1. Partnering and realistic risk allocation in contracts are at one end, and litigation before a judge and jury is at the other. As you move along the continuum, the cost increases, in time, money, and the potential for damaging the parties' relationships. One of the reasons for this is that as you get farther along, you take more and more power away from the parties themselves and place their fate into the hands of a third party. The ideal is for the project parties to work out a mutually satisfactory resolution. To that end, every contract should include the following short but effective dispute resolution clause:

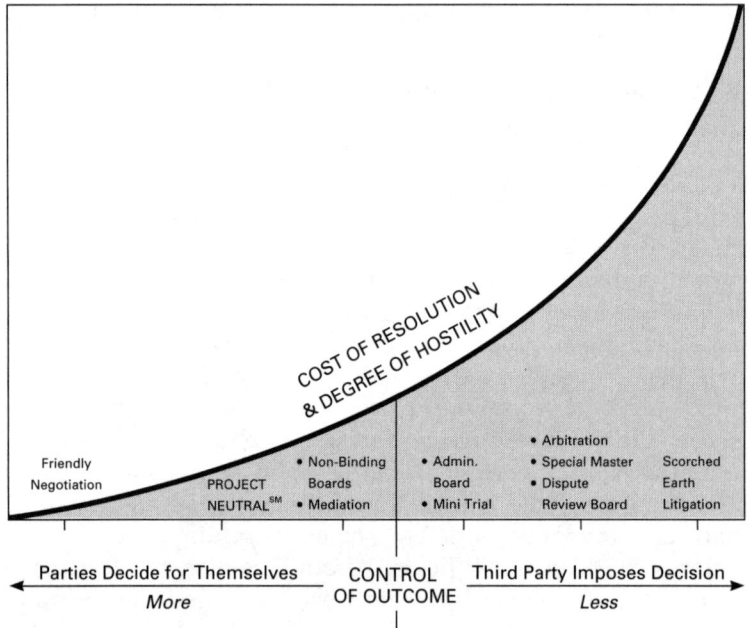

Figure 10.1 There is a spectrum—a range of choices—in approaches to resolving a construction disagreement. At one end is partnering—friendly negotiation. At the other is "scorched earth litigation." Coauthor Rick Steen's company, Hill International, prepared this "Dispute Resolution Continuum" to show the place of the "Project Neutral," Hill's innovation, which is one of the newest ADR options (see pp. 156–160).

> In the event of any dispute under this contract and before taking any other action, each party agrees to appoint a senior representative with authority to bind the party, who will meet personally with the other party(ies) and attempt in good faith to resolve the dispute. [Partnering is a refinement of this, which begins with the two parties' managers at the lowest level (ed.).]

Failing that, it may be necessary to introduce others into the process. The following section is a discussion of several techniques in which the ultimate power to resolve the dispute remains with the parties. Lastly, we will examine a variation on traditional arbitration.

As you read about the various ADR procedures, keep in mind that every dispute is unique, and that the dispute resolution procedure should be designed to fit the dispute.

Step negotiations

It is not uncommon, in the midst of preparation for a trial, to discover that a litigation that has consumed hundreds of thousands of dollars and taken several years to reach the point of trial may have started with a relatively minor disagreement on a job site between a foreman and an inspector which was ignored. That unresolved problem led to other problems, some of which were addressed but most of which were not, and all of which contributed to a delayed project which cost far more than anyone had anticipated and which will be known not for the building that resulted but for the litigation it spawned.

Why do minor disputes go unresolved and mushroom into problems which overwhelm even the most experienced contractors, owners, and design professionals? The reason may lie in a lack of personnel trained in dispute resolution techniques. One of the most important skills in dispute resolution is knowing when a dispute cannot be resolved at your level and must be kicked up the management chain.

Step negotiations, which are an integral part of a partnered contract, involve training all levels of decision makers in basic dispute resolution techniques such as listening and communication skills. In addition, decision makers should be trained to recognize the importance of resolving disputes as soon as they arise and to know when the resolution of a dispute is beyond their capability or authority. Once this happens they are required to push the disagreement up to the next level of management. Under step negotiations, disputes are constantly monitored by more senior levels of management for as long as the dispute remains unresolved.

Dispute review boards

Dispute review boards (DRBs) were first used in tunneling contracts, but in recent years they have been used on a variety of construction projects. While there is a perception that DRBs are suitable only to very large projects because of their cost, this need not be the case, since you can easily tailor the cost of the DRB to fit the project.

The DRB is established in the contract. The board, normally comprised of three members experienced in the type of construction involved, are selected jointly by the owner and the contractor or, in cases of publicly bid contracts, appointed by the public body. The role of the DRB is to monitor construction, keeping abreast of progress and any potential problems.

Since the DRB members are involved from the beginning, they not only know the parameters of the project but also the parties involved. The DRB monitoring may involve actually attending job meetings or simply reviewing project documents on a periodic basis.

Once a party has identified a dispute which the parties have been unable to resolve themselves, they ask the DRB to convene a meeting. Since the DRB has been monitoring the project, the presentation should be relatively short and can focus solely on the issue in dispute. While the meeting is generally informal, the more comprehensive and detailed information the parties present to the DRB, the better able the DRB will be to render a decision.

Under the typical DRB clause, the board may take any one of several actions after the parties have presented their sides. The DRB may ask one or both parties for further information. They may make a decision on entitlement, for example, rule that the contractor is entitled to an extra, and suggest that the parties negotiate the amount to be paid.

While the DRB's decision is not binding, it is admissible as evidence in any subsequent proceeding. Since the DRB's decision is made by experienced individuals close in time to when the dispute arose, it will usually be given great weight by arbitrators, judges, and juries. This feature of DRBs—the admissibility of DRB decisions—is intended to encourage the parties to accept the DRB's decisions, allowing the project to proceed without lingering disputes.

The DRB is only an advisory body but, if the parties heed the advice of the DRB, the project can proceed without too much time and money being spent on disputes.

Case study: co-generation facility dispute review board. The contract between the owner of the co-generation facility and the contractor contained a detailed dispute review board clause, much like the procedure described above.

Early in the project, the contractor put the owner on notice that, due to changes dictated by the owner, the contractor and its subcontractors were facing increased costs of performance. The owner requested documentation to support the claimed increases, but did not receive all of the information it requested and therefore decided to take no action on the claim. The contractor then advised the owner that without at least an interim payment, the contractor and its subcontractors would be unable to proceed with the work.

After several months without resolution, the owner and the contractor agreed to submit the dispute to the dispute review board at the next regularly scheduled job meeting, which was in two days. The next day, the DRB panel was notified that a dispute would be submitted at the job meeting on the following day.

At the DRB hearing, both parties submitted position papers and supporting documentation to the DRB. Several days later, the contractor sent additional documentation to the DRB. Within a week of the initial hearing, the DRB held a private meeting, and the next day it issued its recommendations to the parties. These recommendations included having the owner make an interim payment to the contractor and the contractor submit proper documentation within 30 days or returning the payment. The DRB also gave the owner and the contractor suggested parameters for pricing future extra work claims.

In addition to a recommendation on the dispute before it, the DRB told the parties that their use of the DRB procedure in this instance was not as effective as it could have been. The DRB noted that while the parties had been aware of the dispute for some time, the DRB was given only one day's notice of a hearing and did not have adequate time to prepare.

The presentations at the hearing were considered to be unprofessional and confusing and required the DRB to spend more time than necessary in deciphering the parties' positions. In the future, the DRB recommended that, for each claim, the contractor should identify each item of dispute, set forth the grounds for entitlement, and establish a dollar value. The owner should then respond to each item, discussing entitlement and dollar value.

The lesson here is that while DRBs, like other forms of ADR, are intended to be informal and to save time and money by eliminating some of the procedures used in litigation and arbitration, there is still a need for organized, detailed, and coherent presentations to the neutral.

Mediation

Mediation is perhaps the most widely talked about and most commonly misunderstood method of dispute resolution. The most common misconception is that mediators settle disputes. They don't. Parties settle a dispute as soon as they decide it is in their mutual interest to do so. At the most basic level, the mediator's function is to keep the parties talking and searching for ways to resolve the dispute. Mediation is a structured negotiation in which the mediator provides the structure. The mediator establishes ground rules and acts as a referee, facilitating communications between the parties. The following case histories demonstrate how and why mediation can be successful in resolving construction disputes.

Case history: corporate headquarters project. This project involved the development of a new regional headquarters for a worldwide company in a growing market area. A construction manager, which had performed other work for the company, was asked to help locate an appropriate building and find an architect for the project. There was an assumption on the construction manager's part that he would be awarded the contract to build the project.

As the project planning proceeded, the construction manager submitted requests for reimbursement of its costs to the company, which were paid. The arrangement was informal, and there was no document setting forth the basis under which the construction manager would be compensated for its efforts, nor exactly what those efforts were to be.

When the design was essentially complete, building permits had been obtained, and demolition work started, the owner's home office became concerned about the scope and cost of the project and the lack of a formal contractual relationship. When a contract acceptable to the home office could not be negotiated, the construction manager was directed to cease all activity related to the project.

When the construction manager submitted its final invoice, the owner refused to pay, contending that it had overpaid the construction manager on past invoices; the construction manager's costs were inflated, duplicative, and in certain cases fraudulent; and the construction manager's mishandling of the project had delayed its start and increased the project's cost.

In addition to canceling the headquarters project, the company canceled all other contracts on which the construction manager was working and stopped all payments to the construction manager. The company had represented a substantial portion of the construction manager's volume. In addition, word of the abrupt end of the relationship between the company and the construction manager had reached some of the construction manager's other clients and was affecting its ability to obtain new work.

After exploratory settlement discussions, it became apparent to both the company and the construction manager that litigating the dispute would be not only costly but also risky for both parties. There was even a dispute as to where the litigation should take place and what law should apply, that of the United States or the foreign country in which the headquarters was to be built. Since there was no written contract, the case would turn largely on the memory and credibility of each party's witnesses.

Instead of litigating, the parties agreed to mediate the dispute and retained a professional mediator to assist them. The mediation took place over two, 10-hour days and resulted in a settlement under which the construction manager was paid all of its documented costs, plus overhead and profit. In addition, it was paid all amounts due for work under other contracts, as well as some termination expenses. The construction manager also received a letter from the company president praising its long relationship with the company and the company's happiness with the quality of its work. The company was able to devote its attention to the renovation of its new regional headquarters, not litigation.

Case history: homeowner/contractor mediation. This matter involved a claim by homeowners that work at a construction project adjacent to their home not only damaged the house but led to one of the homeowners contracting colitis, a disease which is often related to stress. The contention was that the developer and the construction manager misled the homeowners about the impact the project would have on them, and intentionally failed to tell the homeowners when blasting activity would take place. The homeowners claimed that they were in a constant state of worry about their home and their personal safety. Portions of the house had been constructed prior to the Revolutionary War and it had been designated a historic structure.

The defendants included the project's developer, the construction manager, the site work contractor, a subcontractor who was responsible for the blasting operation, and the trucker who hauled excavated material from the site. The construction manager, a large national outfit, was the plaintiff's main "target," and the only party against whom they sought punitive damages.

It was estimated that a trial would take not less than three weeks and would involve an unusually heavy cost to all parties for expert witnesses. It was anticipated that between the plaintiffs and defendants there would be experts testifying in the fields of historic structures, structural engineering, blasting, heavy construction, trucking, and stress-related diseases. The case was settled after a one-and-a-half-day mediation for less than what the trial costs would have been.

The mediation process. A major function of the mediator is to keep the parties talking and exchanging information. It is information (usually the lack of it or misunderstanding of it) that causes disputes.

One way a mediator keeps the parties talking is by acting as a sounding board for the parties in private caucuses. It is not uncommon for negotiations to break down when one side insists on interpreting the events leading up to the dispute in a way that the other side absolutely rejects. While the parties may agree on the facts, they disagree about the implications of those facts. For example, in the headquarters project mediation, any time the construction manager mentioned the "contract," the company's representatives interrupted to say that there was no contract and they objected to the construction manager using the term. Of course, the construction manager insisted that he in fact had a contract, and perhaps believed that if he repeated the word often enough, the owner would eventually agree.

In a private caucus, the mediator pointed out to the construction manager that as soon as he said the word "contract," the company's representatives stopped listening to him. The mediator suggested that in the next joint session the contractor omit any reference to the "contract" and simply describe what had happened and why he was seeking the money he was. The next joint session turned out to be a very productive exchange of information about financial data, which eventually paved the way for framing the settlement.

Mediators can also be used to test the impact of what one side believes is critical evidence, but which they do not want to disclose prematurely to the other side. In the headquarters project mediation, the construction manager had a copy of a letter from the owner to a third party which seemed to acknowledge the existence of a contract. In a caucus with the mediator, when the mediator asked why the construction manager was so certain he had an enforceable contract, the contractor showed the letter to the mediator. The contractor was surprised to learn that in the mediator's caucus with the company's representatives, the letter had been brought up and discussed in detail, and the company believed they could explain why their position that there was no contract was not inconsistent with the letter. The construction manager was then forced to reevaluate his position, since it appeared his "smoking gun" was shooting blanks.

One of the biggest stumbling blocks to resolving disputes is distrust and suspicion. The result is that any suggestion by one side is viewed with suspicion by the other side, frustrating the search for a mutually satisfactory resolution. One way around this dilemma is for the parties to work on what caused the breakdown and attempt to reestablish some of the trust and respect that had at one time existed.

If the relationship can be reestablished, not only may the dispute which led to the mediation be resolved, the parties may be able to work together in the future. Unfortunately, this was not the final result in the headquarters project mediation.

However, the mediator was able to help the parties get past the mistrust and hostility enough to allow the company and the construction manager to resolve the dispute. The breakthrough which led to the settlement came when the owner's representatives began honestly to evaluate the construction manager's contention that the company was overstating the cost to complete the project. Not only was the construction manager right, the company could save additional money if it used a major material supplier the construction manager had planned on using.

When the company offered to pay the construction manager the "savings" and the construction manager agreed to contact the supplier and ask him to give the owner the same terms it had quoted to the construction manager, the matter was settled.

Advantage of mediation: parties speak for themselves. You will often hear parties to a dispute say that if they simply got an opportunity to "tell their story," any judge, jury, or arbitrator would rule in their favor. Unfortunately, you rarely get such an opportunity. In a trial, and in most arbitrations, your "story" will be told through responses to your attorney's questions, and the storyline will often be interrupted by objections from the opposing attorney. It is not unlike someone constantly switching channels on the TV. You want to watch one show, and the other party wants to find out what is on the other 38 channels.

After your attorney finishes the direct examination, cross examination begins. During that process the examiner starts with the premise that you do in fact have a story to tell. However, the other side will tell the jury or arbitrator that your story is nothing more than a great fiction. The premise is that the truth will survive this adversarial battle. It may, but you and your wallet may not.

In a mediation you will get a chance to tell your story, but not to a jury or an arbitrator. Rather, you will tell your story to the other side, which, for any number of reasons, probably never heard it before, or did not understand it. This feature of mediation is one of the factors that led to a settlement in the homeowners/contractor mediation.

One of the homeowners' major complaints was that the contractors had ignored their concerns before and during construction. They felt that no one was listening to them. At the mediation, they finally got the opportunity to tell the contractors what they were going through and why what seemed not to be serious damage was a concern because of the age and historic significance of the house.

Not only did they tell their story, they also heard the construction manager's story. They learned that the construction manager had taken extensive steps to minimize the construction's effects on the homeowners, and that in many instances it had done more than the law required. Unfortunately, until this point the homeowners had been unaware of these efforts because there had been no real communication between the homeowners and the construction manager. As it turned out and as the construction manager admitted at the mediation, the construction manager's failure to keep the homeowners informed had led to the perception, if not the reality, that the construction manager simply did not care how the construction affected the homeowners.

Once the homeowners understood that the construction manager had in fact tried to minimize any damage and inconvenience, the tenor of the case changed. Instead of seeking hundreds of thousands of dollars for "pain and suffering," plus punitive damages, the homeowners were willing to accept compensation for their actual damages. And the construction manager, who had faced the largest claim, ended up contributing a very small amount to a settlement funded in large part by the excavator and the blasting contractor, whose activities had actually caused the damage.

One of a mediator's most valuable functions is making sure that the agreements reached in the mediation session are carried out. Normally, upon the successful conclusion of a mediation, the mediator prepares a memorandum of understanding, outlining the settlement and describing the steps that must be taken to finalize it, such as the preparation of formal settlement documents, releases, etc. The memorandum of understanding should be relatively short, clear, and concise.

A twist in the corporate headquarters case was the appointment of the mediator as the sole and final arbitrator of any and all disputes arising out of the memorandum of understanding. The appointment came about when the mechanics of preparing a settlement agreement almost caused the agreement to come apart at the last minute.

Toward the end of the second day of the mediation, when many of the issues had been resolved, it became necessary to call in an additional attorney for the parent company to "sign off" on certain aspects of the settlement. Not having been involved in the mediation, this attorney was unaware of much of the thinking that had gone into the settlement agreement and was concerned that the memorandum of understanding was rather vague and imprecise.

He was right, but during the course of the mediation the parties had reestablished a certain amount of trust and were convinced that each knew its rights and responsibilities and would be able to carry out actions required by the settlement without problems. Since it was going to be physically impossible to prepare a detailed settlement agreement that evening, and because the parties did not want to end the session without some sort of agreement, the parties approached the mediator and asked if he would deal in a summary fashion, as an arbitrator, with any issues that arose during the implementation of the settlement. The mediator agreed and the memorandum of understanding was signed.

While the memorandum spelled out tasks to be performed by each party over several months, there was one more problem, which involved something neither side had considered. That problem was resolved in a telephone call with the mediator that lasted less than a half-hour.

The minitrial

The minitrial is a condensed version of a typical trial or arbitration and may be compared to an executive summary of a lengthy report. The idea is to present as much information as possible in as short a time as possible to someone who is not directly connected with the dispute. This individual then evaluates the information and gives the parties his or her assessment of the dispute in the form of a nonbinding advisory opinion.

If the third party is a neutral, his or her role generally ends at this point. If the presentation is made to a panel consisting of a neutral and a representative of each party, the three will use the advisory opinion as a basis for further negotiations. In the following case histories, a hospital project dispute led to a presentation to a single neutral and served as an alternative to a trial, and an industrial project involved, as an alternative to arbitration, a presentation to a panel of a neutral and two party representatives.

Case history: hospital minitrial. This project involved the construction of a $200 million hospital for a not-for-profit foundation. Completion was delayed for almost two years largely as the result of disputes over extra work and contractor claims. Numerous lawsuits were eventually consolidated into one case involving nine parties. At the suggestion of the judge handling the case, who was concerned about the length of a trial of this very complex matter, the parties agreed to engage in a "minitrial" which would be chaired by a neutral advisor selected by the parties.

With the assistance of the advisor, the parties agreed that everything said or done in connection with the minitrial would be confidential and off the record. The minitrial was scheduled for a full week from a Monday morning to a Sunday afternoon. Each party was permitted four to six hours to present its case and could do so in any manner it chose. Some used panels of witnesses, others used charts and diagrams, and still others presented witnesses with a question-and-answer format. There was no cross examination, but the parties could suggest questions to the neutral advisor, who could, but was under no obligation to, ask those questions. The schedule was generally one presentation in the morning and one in the afternoon.

After a group dinner each evening, the neutral advisor convened the parties for what he called "confrontation sessions." The neutral advisor chaired these from the front of a large amphitheater, with witnesses and party representatives in the front rows. Lawyers sat in the back and were not permitted to speak. The neutral advisor used the sessions to ask questions and to urge the parties to address contradictions and conflicts in the presentations.

At the conclusion of the minitrial, the neutral advisor prepared a nonbinding decision allocating responsibility for the problems. While that decision did not lead immediately to a resolution of the dispute, it was used as a framework for settlement negotiations, and eventually six of the nine parties settled.

The parties that did not settle spent five months trying the case. The result was a jury award in excess of $26 million, which was later overturned on appeal.

Case history: prearbitration minitrial. In an attempt to settle a dispute arising out of changes and delays in the construction of an industrial facility, and as an alternative to the binding arbitration called for in the contract, the owner and designer/builder agreed to participate in a minitrial.

Under the terms of the minitrial agreement, each party agreed to have a management representative attend the minitrial and participate in follow-up negotiations. The representatives, each of whom had the authority to settle the case on behalf of their party, agreed to listen to all the presentations and to discuss in good faith the relative positions and to seek a resolution of the differences.

In addition, the parties agreed on a neutral advisor whose role was to advise each party of his opinion of the likely outcome of the arbitration, should it proceed. The advisor was also responsible for moderating the presentations and keeping the proceedings moving. All discussions with the neutral were to be in the presence of both parties.

The minitrial agreement provided for limited pretrial discovery and for any discovery disputes to be submitted to one of the arbitrators for the matter, who also had the authority to impose any sanctions he deemed appropriate regarding discovery disputes. The parties also agreed to exchange all exhibits to be used at the minitrial, a detailed cost analysis, and a position paper of no more than 50 pages from each side.

At the minitrial, each party was given a limited time in which to present its case, and all interruptions were prohibited, including questions by the neutral advisor or the party representatives. All questions were reserved for a specified question-and-answer period following the presentations.

At the conclusion of both presentations, the management representatives were to meet and attempt to resolve the dispute. At the request of either, the neutral advisor could render an advisory opinion, making recommended findings of fact and law. Following his oral advisory opinion, the management representatives were to meet again in an attempt to settle the dispute. If there was no settlement, the neutral advisor was to render a written opinion.

In order to induce the parties to resolve the case without arbitration, the minitrial agreement provided that either party which rejected a last written offer of the other that was reasonably similar to the arbitrator's award would be responsible for all arbitration and minitrial costs, including attorneys' fees and arbitration fees.

An award which was within 10% of the last written offer was considered reasonably similar. If the designer/builder offered to accept $1.5 million, the owner rejected this offer, and the award was then not less than $1.35 million, the award was considered similar to the demand and the owner would be liable for sanctions. On the other hand, if the owner's offer was $250,000, it was rejected by the designer/builder, and the award was $275,000 or 110% of the last offer, then the designer/builder would be liable for sanctions.

While the minitrial procedure did not result in an immediate settlement, the case was settled soon after the arbitration hearings began. The parties credited the minitrial process as a significant factor in the settlement.

Binding arbitration

Arbitration is what the parties make it. It can be costly, time consuming, and lead to absurd results, which are for all practical purposes unappealable.

While arbitration is created by the contract, it is possible for an arbitrator's award actually to ignore expressed contract terms and leave the parties unable to do anything about it. On the other hand, arbitration is seen as a way to assure equity and justice without the cost and time associated with court proceedings. It is seen as an informal procedure where the disputants get to tell their stories to a neutral party experienced in the industry, who will render a fair decision in a relatively short period of time.

Arbitration can be a very effective form of dispute resolution if used properly. Unfortunately, parties typically agree to arbitration at contract signing, well before any disputes arise, and generally with the hope and expectation that there will be no disputes. Once a dispute arises, whether it is appropriate or not for arbitration becomes academic if either side wishes to enforce the arbitration clause. *The best time to select a dispute resolution method is when the issues are known and the method can be designed to fit the dispute.* Arbitration can easily be tailored to fit a particular dispute.

Baseball arbitration

One form of arbitration which can be used only in certain types of disputes is known as "baseball arbitration," the name coming from the model of salary arbitration called for in the Major League Baseball collective-bargaining agreement. Under this form of arbitration, each party states its last best offer in writing. The arbitrator then has limited authority to render an award, selecting either one figure or the other as the more reasonable. The arbitrator may not make a compromise award.

The effect is twofold. First, the parties have a strong incentive to be reasonable, and settlements short of arbitration are encouraged. Second, the time required for arbitration is reduced, because the arbitrator need not be presented with all the evidence necessary to reach an independent evaluation. The sole issue is which offer is more reasonable under the circumstances.

Case history: hydro project baseball arbitration. An owner and a contractor signed a letter of intent to enter into a construction contract for a hydroelectric project. The contractor was authorized to proceed with certain preliminary work, but with the understanding that if financing was not obtained the owner had no obligation to award the contract and the contractor would be entitled to be paid only up to the maximum amount set forth in the letter of intent.

Unfortunately, the letter of intent did not fully detail exactly how amounts due the contractor would be calculated. When the owner could not get its financing in place and the project was canceled, it turned out that the owner and the contractor had widely divergent views as to the value of the work performed, much of which had been mobilization and preparatory work which the owner, if it eventually did go forward, would have to repeat.

Using a baseball arbitration format, the parties selected a single arbitrator and scheduled a finite time for each side to present its case. They then exchanged their last best offers. When they still could not settle the dispute, they proceeded to arbitration.

At arbitration, the contractor was given a single day in which to present facts and circumstances justifying its number, as was the owner. Each side then had a half-day to rebut the other side's presentation.

The arbitrator was then given seven days to select either the contractor's number or the owner's number as more reasonable. The arbitrator selected the contractor's number, which was $400,000 higher than the owner's number. When the dispute first arose, the contractor was claiming $1.5 million, while the owner claimed the owner was due $250,000. (Note that the difference in the size of the two parties' figures dropped by about two-thirds thanks to baseball arbitration.)

Baseball arbitration can also be easily applied to extra work claims where there is no dispute that the work is extra, the issue being the value of the extra work or credit.

The State of the Art in Dispute Resolution

In the past few years the construction industry has embraced ADR for resolution of disputes. Partnering is a recognition at the outset of a project that the parties are willing and committed to work together to achieve mutual objectives. It comes as no surprise that these two complementary concepts have grown in parallel.

The key objectives of parties to a dispute are to manage and control the process of resolving the dispute, understand the validity of both positions, maintain control of the outcome, and prevent future disputes.

The construction industry is just now realizing that neutral expertise, on hand during the project, is a vital component in understanding the parties' positions and in resolving disputes. The emergence of the role of a neutral expert is one of the newest advances in successful project dispute resolution. Hill International, Inc., a world leader in construction claims management and dispute resolution services, has established "Project Neutral" as an advanced dispute resolution technique that meets all of these objectives.

The role of the Project Neutral in dispute management is the last link in the chain of comprehensive, cost-effective dispute management. The concept combines the advantages of neutral expertise with those of consensual ADR techniques that are less formal than arbitration. It is a logical extension of dispute management theory, and is being used on a wide variety of projects.

The Project Neutral offers numerous advantages over arbitration alone, including more control by the parties of the process and of the outcome, and preservation of relationships. Among advantages are the following.

- All too often, because of the complexity and uniqueness of each construction project, a sole mediator may not possess the requisite knowledge in the specific subject matter of the dispute to provide the parties with the technical expertise necessary to assess the facts of the dispute and inform the parties of the soundness of their positions. Engaging a firm that has a broad spectrum of technical expertise and staff skilled in the dispute resolution process is the preferred solution.
- There are limitations to having each party bring in an expert. In many cases, dispute resolution techniques that do not provide for neutral fact finding and neutral expertise require each party to engage experts and/or divert limited project personnel resources to document facts and issues in the dispute, while sacrificing continued satisfactory project progress in critical areas of the project that are not affected by the dispute. When each party retains an expert, the frequently conflicting factual histories and opinions of those experts adds a degree of adversity, may harden the parties' positions, and adds to the time and cost of resolution.

The Project Neutral provides a wider range of dispute management services with a substantive focus that offers clear advantages over other techniques used in recent years:

- The Project Neutral combines job-site facilitation services with independent and neutral fact finding to give the parties a fair and objective assessment of the merits of their positions and to facilitate the settlement of the dispute.
- The Project Neutral allows the parties to maintain maximum control, to benefit from the neutral firm's expertise in helping the parties to understand their positions, and to resolve the dispute while minimizing its impact on the project.

Because of its success, the popularity of the Project Neutral is growing.

All of the advantages of ADR techniques are available through the Project Neutral. As discussed earlier, the timeliness of resolution is key. When the Project Neutral is engaged at the outset of a project, the familiarity with the project gained by the Project Neutral enhances the timeliness of the assessment of the technical and factual merits of the positions of the parties. Providing for a single participant on a continuing basis, who has access to experts in the issues in dispute, has clear advantages compared to having a board monitor the project.

One element related directly to the time needed for resolution is the cost of the process. A Project Neutral with full and open access to all of the personnel and records of all of the parties to the project has the ability to cut through the positioning and posturing that always exists in disputes, and get to the facts. In the litigation and arbitration of major disputes, it is the rule rather than the exception that, in addition to the battle of the attorneys and the battle of the parties, the process is further burdened with the expense of the battle of the experts.

At the outset, the Project Neutral is an informal process, with the parties controlling the degree of formality, the degree of confidentiality, and the use of the outcome as precedent. This informality reduces the cost of the process.

There is a strong synergy between appropriate and effective advocacy by counsel and authoritative neutral expertise, in getting to the reality of the parties' positions and allowing the resolution process to move forward. One of the advantages of traditional mediation is to create a win/win situation, where both parties come out of the dispute satisfied (although not always pleased) with the outcome. An added advantage of the Project Neutral is in giving all parties a clear and rational basis, under the contract and under the facts, for deciding how to resolve their dispute. This adds measurably to the parties' perceptions that the outcome is fair and equitable. It also provides a factual record and justification for settlement, contract changes, and funding authorization.

The Project Neutral is a combination of dispute resolution boards and partnering. The Project Neutral is particularly effective in partnered projects as an ongoing sounding board for potential disputes, a neutral entity that can keep the parties focused on the partnering goals, and one that can keep the day-to-day stresses and problems on the job from impeding open communications and burdening the overall project progress.

A typical provision in a construction contract for engaging the Project Neutral at the beginning of a project covers the following items:

- Designate the firm engaged to serve as the Project Neutral.
- Identify the scope of initial familiarization and ongoing involvement with the project.
- Specify when in the dispute resolution procedure a party can call upon the Project Neutral for assistance.
- Provide that the parties share the cost of the Project Neutral equally, and identify under what circumstances the Project Neutral can assess costs against one party.
- Indicate the authority of the Project Neutral while serving in such role(s) as the parties provide, including facilitator, fact finder, and/or arbitrator.
- Specify the schedule and procedure for fact finding, presentation of preliminary findings, rebuttal, and final results.

Typical provisions in a separate agreement, after dispute has arisen, for engaging the services of the Project Neutral include:

- Recitation of the contract and the dispute procedural setting.
- Recitation of the voluntary agreement of the parties to utilize the process—to resolve disputes without lengthy or costly arbitration or trial proceedings.
- Identification of the principals and other representatives, for purposes of points of contact, settlement negotiations, and their authority.
- A statement that the process is without prejudice to the parties' rights to institute other contractual or legal proceedings.
- Provision that one party, on behalf of both parties, contracts for the Project Neutral agreed upon by both parties.
- Definition of the scope of work for the Project Neutral.
- A statement that the Project Neutral shall be free to contact any person and review any file or document, including "confidential" files.

- A provision that the Project Neutral fees shall be borne equally by both parties, or as otherwise determined.
- A provision that all costs to a party, except the Project Neutral fees, shall be borne by that party.
- Agreement that the Project Neutral and the parties shall treat the process as confidential and undertaken for purposes of settlement negotiation, and that no aspect shall be disclosed to third parties.
- The Project Neutral is disqualified as a witness, consultant, or expert for either party in any subsequent proceeding involving this dispute.
- The Project Neutral work product shall be considered protected and prepared for purposes of settlement negotiations, inadmissible as evidence, and shall not be used in any subsequent action involving this dispute.

The advantages of the Project Neutral over other forms of consensual or third-party imposed dispute resolution outcomes are clear. The Project Neutral is a partnering facilitator with two advantages: He or she is (1) continually on tap, and (2) technically knowledgeable. If they wish, the parties can use the Project Neutral to enhance their future operations and relationships and avoid or prevent similar disputes from occurring.

Project Neutrals at Work

Two examples of the successful utilization of the Project Neutral in helping the parties reach agreement, while preserving their relationship, are described below.

In the first instance, a state-of-the-art aquarium project was contracted with a provision for a designated neutral, the Project Neutral. During the course of the project, a number of issues of interpretation of the contract requirements arose. The Project Neutral was successful in evaluating the facts and the contractual basis for the positions of the parties, advising the parties of the strengths and weaknesses of their positions, and bringing the parties together to achieve a consensual resolution of the issues. As important, the process allowed the parties to agree on interpretations of requirements for other aspects of the remaining project effort, and to prevent those issues from becoming disputes.

In another successful application of the Project Neutral, a federal agency and a contractor agreed to use the Project Neutral to help resolve a substantial claim on a project for modernizing communication facilities in several different countries. The services of the Project

Neutral included evaluation of the contractual requirements, detailed fact finding, and presentation of expert opinions on the strengths and weaknesses of the parties' positions, as well as on the likely outcome of a binding third-party decision. The fact finding was undertaken through interviews of project personnel and review of project documents. Preliminary findings were presented to the parties and their counsels, and the parties were permitted to provide supplementary information while the Project Neutral team finalized its findings. The Project Neutral then facilitated the negotiations between the parties, which concluded in an amicable settlement.

Conclusion

Partnering is a recognition that disputes can and do arise, even among the most well intentioned contracting parties. Reasonable people dealing in good faith can disagree. Partnering and ADR are tools which allow the parties to resolve those disputes without destroying the value of the contracts that brought them together. They are proactive in that they anticipate potential problems and seek to resolve them through win/win solutions.

Attorneys are hired to protect their clients' interests, and they have an obligation to do so. However, they may lose sight of the fact that the client's paramount interest in a contract is a successful business relationship. Attorneys should not draft contracts and dispute resolution clauses solely to "protect" clients in a case of default. Instead, they should draft contracts and dispute resolution clauses to advance the object of the contract: a successful completion of the contract with a fair profit for all.

Part 3

Nonproject Uses of Partnering

Chapter 11

Steel Erectors and OSHA Partner to Dramatically Improve Safety

Steve Miller

The term partnering is most often used to denote a project management tool that includes a preproject workshop, but it also has other applications—such as contractor–worker partnering in the name of safety.

After a June 1992 Occupational Safety and Health Administration (OSHA) decree that steel-erection safety rules in the Mountain States would be tightened radically, all Colorado steel erectors gathered to decide what to do. They agreed to share with each other all they knew about safety in steel erection, from written policies to fall-protection equipment. There were to be no trade secrets when it came to safety. Everyone agreed that all steel-erection job sites would be open for inspection by any other member of the group, for purposes of providing insight into safety practices.

The steel erectors also got OSHA Region VIII (Denver) Administrator Bart Chadwick to visit an ongoing project, where he watched iron workers at work. OSHA had prohibited "connectors" (who bolt beams to columns) from walking across a beam when only one end is bolted and the other is held up by a crane. The worker had to descend the column, walk over to the other column, and scale it.

That is, until Miller asked OSHA's Chadwick to strap on an iron worker's safety belt and climb a column. Soon, OSHA's policy was changed to allow crossing these "live loads," provided the connector had a positive means of fall protection attached to the beam before *crossing it. (Editor's note: Author Miller created such a positive fall-protection system; see Fig. 11.1.)*

Since the creation of the Occupational Safety and Health Administration (OSHA) in 1970, steel erection safety has been one of the most controversial and hotly debated issues in construction. Subpart R, the standard that covers steel erection, was a poorly conceived hodgepodge of rules and regulations that were gathered from whatever written rules existed at the time. Little research went into determining what were accepted industry practices or erection procedures. Instead, most of Subpart R referenced work practices that were outdated years before the standard was written. For example, riveting operations have been replaced by high-strength bolts or welded connections on most buildings today, and had been long before 1970, but were described in the standards. The standard calls for gathering and stacking wooden planks on the erection or derrick floor. Wooden planks were replaced by light-gauge metal decking 30 years ago or more.

Falls from heights are the most common cause of injury and death in the steel erection industry. Not surprisingly, most OSHA inspectors focus on fall hazards when conducting compliance inspections. And because so few OSHA compliance officers have any practical experience in steel erection, their perception of danger is quite different from that of ironworkers who work at heights every day.

While all OSHA inspectors may not be well versed in steel erection safety from either a practical standpoint or that of the letter of the safety law, they are smart enough to know that if an ironworker falls 30 ft to a concrete floor, the worker is going to be seriously injured or killed. Yet the standard allowed for falls of up to 30 feet, and does today. (A new standard was being drafted as this book was going to print.)

So all inspectors knew that the language in Subpart R would allow people to be killed in the course of their work, but their hands were tied. In an effort to reduce these exposures, OSHA began citing steel erectors under different standards that did not apply to steel erection. Before long, there were as many different interpretations of steel-erection height requirements as there were OSHA inspectors to enforce them. To make matters worse, OSHA people themselves were as confused as the steel erection community. Rules varied from state to state, and even within local offices.

In the meantime, unwary employers were receiving frivolous citations. And because fines were relatively low when compared to the cost of fighting citations in the courts, the employers paid the fines and went about their business. Unfortunately, this did little or nothing to improve job-site safety practices. What many employers failed to realize was that while OSHA was dealing out fines that were no more than a slap on the wrist monetarily, with each successive citation, a contractor was hit progressively harder.

For example, suppose an employer was cited for allowing (or "instructing," to use OSHA language) employees to use a defective ladder that has a split side rail. The initial fine for this might be $5000, with the possibility of an 80 percent reduction because of the size of the company, their good-faith effort to abate the hazard immediately, and their past history, which up until then had been spotless. The fine could thus end up being only $1000, hardly worth fighting in the courts.

The real damage was the fact that the employer now had a serious citation on his record. If the same employer received a similar citation in the next couple of years, he would receive a "repeat serious," which could lead to the second fine being two or three times as much as the first. A third violation could result in a fine 10 times that of the original. The next step, a "willful violation," has a base fine of $70,000. The final step is "criminal willful." Criminal willfuls are pretty rare but could result in the employer doing time in prison.

One might think that it would be ridiculous for an employer to go to prison over a ladder citation, or even a series of ladder citations, and that person would be right. However, if the employer were unlucky enough to have an employee seriously or fatally injured because of a ladder-related fall, his past record would indicate that he had shown indifference or flagrant violation of the standards in the past and therefore might qualify for a criminal willful.

For 20 years, then, the battle raged between steel erectors and OSHA. The steel erectors claimed they were being wrongfully persecuted, and OSHA claimed they were only trying to protect people from hazards that all too often led to death.

In the early spring of 1992, OSHA Region VIII (Denver) sent out invitations to most of the larger construction trade associations in Colorado to participate in a joint OSHA–construction industry working group. Selected to represent the Colorado chapter of the American Subcontractors Association (ASA) were Doug Hackett, its president, and myself. Doug Hackett, the owner of Hackett (steel) Deck Erection, had a good working knowledge of the OSHA standards. I had spent most of my work life as a steel erector, before becoming a construction safety consultant.

Doug and I both saw this meeting as an opportunity to get the straight story on steel erection safety from safety experts. On the day of the meeting in early April, OSHA had several representatives from Washington, D.C. Also present were several representatives from Region VIII OSHA in Denver, including the regional administrator and four or five of his immediate aides. Colorado has two area directors who head up the compliance division; they were both in attendance.

After several hours of discussion on various matters pertaining to safety and health in the workplace, Doug and I asked a few simple questions about fall protection as it related to steel erection. No one jumped up to answer. In fact, the entire group of OSHA representatives huddled together for several minutes before announcing that they did not have a clear answer for us. With that, we suggested that perhaps we could arrange for another meeting in the future to include only those representatives from the group that had an interest in fall protection. The OSHA people agreed.

The next meeting was held June 10, 1992. We were extremely happy to have the opportunity to get the straight scoop once and for all, no matter how financially painful to contractors, just as long as some consistency in safety enforcement practices could be established.

There was such an overwhelming response to the meeting that OSHA had to move the meeting to the federal courthouse across the street from Region VIII headquarters in Denver. On the morning of the June 10 meeting, Bart Chadwick, who is the regional administrator of Region VIII OSHA, opened the meeting. He started out by saying that the OSHA standards were not clear in many cases, and that many standards were antiquated and needed to be updated. He went on to explain that employers had an obligation to protect their people from recognized hazards even when the language in the standards did not require it. OSHA intended to cite employers under the "general duty clause" in these situations. (The general duty clause is a catch-all clause in the federal Walsh-Healy Act that requires employers to protect their employees from all recognized hazards in the workplace.)

Chadwick went on to explain that employers needed to implement what he called "100 percent fall protection" as an element of their safety programs. He was convinced that preplanning, hazard analysis and site-specific safety plans would save lives. Several employers and safety professionals in the group, including myself, agreed for the most part with what he was saying. Chadwick finished his presentation by saying that he fully realized that certain industries were inherently dangerous and that we could probably never completely eliminate all of the hazards associated with trades such as steel erection. We must take action, though, when it is "technologically feasible," to protect our people from falls. He then turned the meeting over to Ed Kassak, OSHA's Assistant Regional Administrator, Federal State Operations. The crowd of steel erectors did not like what Kassak had to offer. He recited three pages of new OSHA enforcement policies with respect to scaffold, steel erection, and precast concrete erection.

This chapter describes interactions between OSHA and the steel erection community. However, it is worth noting that the other two parties involved, the scaffold manufacturers and users, and the precast concrete erectors, took a radically different approach to the problem than the steel erectors, with totally different results. These groups decided to challenge OSHA, to make them prove their point in court if that's what it took. These contractors had a valid defense in that the general duty clause may not be cited when there is a specific standard that addresses the hazard. Clearly, there were specific standards in each of the respective areas that were addressed in the June 10 letter. Also, the June 10 letter came from Region VIII OSHA, not from Washington, D.C. These contractors were challenging the authority of a regional administrator to set policy without going through the national office.

OSHA, on the other hand, had precedent on its side. OSHA had recently won a court case in spite of the fact that the employer had proven that it was within OSHA standards where employees were being exposed to chemicals in the workplace. When employees first began complaining about the effects of overexposure, the employer ran tests to determine the levels of airborne contaminants to which these employees were being exposed. The tests showed that the levels were within the limits set by OSHA on the Material Safety Data Sheet. These are commonly known as PELs or Permissible Exposure Limits. Eventually, an employee died from exposure to the chemical. OSHA investigated the accident and cited the employer under the general duty clause. OSHA maintained that even though the exposure levels were under OSHA's PEL, the employer should have taken stronger preventive measures to protect its employees after discovering that the PEL was not providing sufficient protection.

The June 10 letter affected the scaffold users probably the least of the three groups. It required employers to provide fall protection for employees during the erection and dismantling of scaffolds 10 ft or higher. OSHA said that when it was "technologically feasible" to provide fall protection for employees, that it must be done. Technologically feasible was a pretty broad statement in that it did not imply that fall protection need be provided in all cases, only that it be provided when there was a practical anchorage point. If a scaffold was being erected or dismantled next to a building or structure, a suitable anchorage could be provided. The scaffold manufacturers interpreted the language to mean that employees had to tie off to the scaffold itself in all cases. They argued that scaffold frames in themselves were not strong enough to provide fall protection for an employee during a fall. They went to great lengths to prove their point. They set up a test scaffold and dropped a test weight on the scaffold. The whole thing was recorded on video, showing dramatic results when the scaffold came crashing down under the weight of the load. While this was viewed by some as a major victory for the cause against fall protection, OSHA continued to cite employers under the general duty clause. To my knowledge, no engineering research was done to determine if existing scaffold hardware could be retrofitted to accommodate fall-protection concerns. I believe that three simple ways to strengthen scaffold inexpensively would be to add diagonal cross bracing, add outriggers to the base of the scaffold to broaden its base of support, and to add some sort of counterweight to the base of the scaffold to make it less top heavy and thus less likely to tip over.

The precast concrete erectors were seriously affected by the letter. Basically, OSHA was calling for fall protection during all erection operations at heights above 6 ft. The precast erectors argued that fall protection in the form of tie-off was not practical during erection because of the greater hazard of being crushed by the huge concrete panels that were being swung in by the crane. Some precast erectors' first response to OSHA's new policies was to ignore them. In the months to follow, one precast erector was inspected on two occasions in Colorado. The fines from these two inspections amounted to approximately $90,000. The erector chose to close its doors rather than fight OSHA. This erector still has not resumed operations and does not plan to. Another erector received close to $250,000 in fines after five OSHA visits to its site over six months. It was not long after these fines were handed out that the precast erectors decided to work with OSHA to find new ways to provide fall protection for their employees.

One individual in particular, Greg Gibbons, called my office one day and wondered if I could arrange a meeting with Bart Chadwick. Greg Gibbons is the owner of a successful precast erection company in Colorado. Greg also sits on the safety committee of the Prestressed/Precast Concrete Institute, a well-respected national organization made up of precast erectors and manufacturers. Greg and I met with Bart shortly after that to discuss alternative fall-protection measures that were being implemented in different parts of the country with much success. The meeting was the start of a positive relationship between the precast erectors and OSHA. In fact, the precast erectors followed the lead of the steel erectors by inviting OSHA out to their jobs on several occasions to review the varying safety systems they had developed.

The steel erectors were affected most by the June 10 meeting. Five of the 10 items outlined at the meeting dealt directly with steel erection and three others, indirectly. The three indirect items concerned scaffold, and affected steel erectors who regularly use scaffold on their projects.

1. As with the other two groups, the steel erectors were directed to provide fall protection (commonly a belt or harness around each worker, tied off to the structure) at heights of 6 ft or more above the ground or working surface.
2. The practice of ironworkers traversing a suspended, or even partially suspended, load was prohibited.
3. Finally, the practice of "Christmas treeing" loads was prohibited. Christmas treeing is a slang term for suspending several steel beams vertically over one another.

Immediately after the meeting, Doug Hackett and I set out to arrange for an emergency meeting of all the steel erectors in Colorado. To the best of our knowledge, there were approximately 40 steel erectors in Colorado at the time, and all of them were contacted. These contractors ranged in size from companies with four to five employees to employers with several hundred. At the first meeting, other contractors were invited, including the precast erectors.

Nearly every contractor who was called made it to the meeting to discuss these important issues. The purpose of the meeting was primarily to inform all contractors of the results of the June 10 meeting and to plan some sort of strategy in response to OSHA's new policies. Early in the meeting, the precast erectors decided not to join forces with the steel erectors, using the rationale that the two industries did not have that much in common with respect to the nature of the work and the hazards that went along with it. It was agreed that the group should remain under the auspices of the American Subcontractors Association. A smaller group of steel erectors had been meeting on a monthly basis for quite some time to discuss steel erection in general. The new group, then, was to be an extension of the older one with a few basic changes. The ASA/Colorado Steel Erectors Committee was formally introduced for the purpose of improving safety in the steel erection industry. It was agreed that no other matters would be discussed at the meetings. Because of the seriousness of these new OSHA directives, the group held several meetings a week.

It is important to note that the erectors in this group consisted of both union and nonunion contractors, most of whom competed fiercely with each other. It was quite an accomplishment just to bring all these contractors together in one room.

What was even more incredible was the agreement made by all contractors to share with each other all they knew about safety in steel erection, from written policies to fall-protection equipment. There were to be no secrets when it came to safety. All members agreed that all steel-erection job sites would be open for inspection by any other member of the group for purposes of providing insight in improving safety practices on their own jobs.

The reasoning behind all of this was, in part, that employees regularly move around from employer to employer. The entire workforce is seen as the most valuable natural resource the erectors have collectively. It is in the best interest of all employers to protect their employees as much as possible.

At the second meeting, the committee appointed a smaller group of contractors to act as a steering committee. Rocky Turner and Gus Price were chosen to co-chair the committee for several reasons. Rocky owns a large nonunion company, and Gus owns a large union company. Both companies had good safety records and had much to bring to the table in terms of experience in safety and general knowledge of steel erection practices. It was also decided at this meeting that my firm, Miller Safety Consulting, would provide safety management services to the group.

The committee adopted a plan that outlined how they would address each issue with OSHA. There was no doubt that we would have to agree to disagree on certain issues, but for the most part, everyone agreed that we would do what we could to work with OSHA in a cooperative effort to see if 100 percent fall protection was really feasible. It just made sense that the money spent on defense lawyers and OSHA fines would buy a whole lot of safety equipment. A follow-up meeting was set up with OSHA for July 13, 1992.

In the meantime, the committee kept meeting regularly, standardized safety policies were written and delivered to each erector, and an inspection program was set up to monitor each individual company's progress. The inspection program was conceived in part so that each contractor had some assurances that his competitor was complying with the rest of the group's policies. One of the first reasons given for the lack of safety on job sites was that "I would spend the time and money on safety if my competitor would." By implementing an inspection program administered by a disinterested third party, the contractors could be satisfied that everyone would be given a fair shake.

By the time the July 13 meeting took place with OSHA, we had prepared a list of items that the group had implemented to manage the safety program. We also brought documentation with us that would demonstrate to OSHA that at least some of the items listed in the June 10 letter were not feasible or practical and would actually introduce greater hazards into the workplace than what we started with.

The meeting was attended by Bart Chadwick and three of his assistants from Region VIII and one of the state's two area directors from the compliance division. Six members of the group represented the ASA, including myself, four steel erectors, and one structural steel fabricator. During the meeting, we felt we presented a pretty convincing argument on those issues that we felt were important enough to continue as part of the steel erection process. We agreed that we would participate in an effort to examine the theory of 100 percent fall protection at heights above 6 ft. We also agreed to develop and implement a comprehensive safety program that included written policies on fall protection, training, and self-inspections to include some sort of self-disciplining effort within the group. All of these things had already been implemented by the time this meeting took place.

What we disagreed on was the height at which connectors should have to tie off, leading-edge protection during decking operations, crossing live loads, and Christmas treeing. We decided that the only way to prove that our work practices were safe was to bring the OSHA people out to a job site and show them. Rocky Turner of LPR Construction volunteered to arrange for the meeting at one of his job sites in Denver that was being erected at the time. The building had several stories, so it would provide a good view of most typical steel erection operations at one site. OSHA agreed to the job-site visit with the understanding that the visit was for observation and learning purposes only and not a compliance visit—meaning that no fines would be handed out for questionable work practices.

The visit was viewed as a major success by both OSHA and the steel erectors. A half-dozen OSHA representatives attended the meeting, together with approximately 40 owners of steel erection companies in Colorado. There were several reasons for the success of the visit. The OSHA people came to the job with their eyes and ears open. And we got down to specifics: Several parties of 10 to 20 people each discussed separate safety issues in different parts of the building. The ironworkers had a chance to explain the positive and negative sides of fall protection in a way that OSHA had not heard before.

The best example took place when we were talking to Bart Chadwick on the seventh floor, directly below where the connectors were receiving and tying in the structural steel beams to columns. You will recall that walking across a beam was at that time prohibited until both ends were bolted. When one end of a beam has already been connected to a column, an ironworker walks across it and bolts the beam to the remaining column.

OSHA had prohibited the practice because it was felt that it was dangerous to walk across the beam while one end was supported by a crane. OSHA feared that the crane operator might drop the load, or the ironworker might lose his balance while walking on the beam because of its tendency to wobble until it is securely fastened on both ends.

The alternative means of making the connection would be to climb the first column to make the initial connection, then climb down the column, cross the floor, and climb the other column. To the layperson, I suppose this alternative sounds easy enough. But what no one considered was the physical strain this constant climbing up and down would put on an ironworker.

Columns normally range in height from 10 ft to 30 or 40 ft. At times, the only way to get up a column is to wedge your feet between the column's two flanges and "walk" up. Unfortunately, there is not always a floor for many feet below the work area—nothing to set a ladder or scaffold on. Once a connector goes up on the iron at the beginning of the day, he usually does not come down to the ground except for lunch or to perform some other work activity besides connecting.

While we were explaining all of this to Bart Chadwick, I called the connector down off of the iron and asked him to remove his full body harness, which was loaded down with all of his tools and two bolt bags full of $3/4$-in bolts. A safety belt or harness, fully loaded with tools and bolts like this one, weighs 50 to 80 lb. I took the connector's harness and asked Bart to try it on. After he had, I asked him to try and climb the nearest column. It thus became clear to him that it is not possible to climb up and down columns all day long with this kind of weight on your shoulders.

Soon after that, OSHA Region VIII's policy was adjusted to allow crossing live loads, provided that the connector had a positive means of fall protection. With the advent of horizontal lifeline systems like the one shown in Fig. 11.1, connectors gained the freedom of movement to avoid being crushed by out-of-control loads, yet were still protected from falls.

It is important to note that the issues the contractors were at odds with OSHA over were not violations of existing OSHA standards. These changes in the rules were brought about by OSHA officials, who perceived these work practices as being extremely hazardous. Before entering into the safety profession, I was a crane operator for most of my 20-plus years in construction. I never personally lost control of or dropped a load, nor did I ever witness any other crane operator drop a load while an employee was on it. That is not to say that it never happens, but I do believe that some statistical evaluation should be performed before making a decision that could have such a major impact on the entire industry. This point was stressed to OSHA officials.

The ASA/Colorado steel erectors tried in earnest over the next two years to make 100 percent fall protection work. LPR's system consisted of attaching vertical stanchions at either ends of each main carrying beam with a $3/8$-in wire rope attached between them. The stanchions were offset from the center of the beam so that the ironworker could traverse the beam without interference from the cable. The stan-

Figure 11.1 Now steel workers walking the beam on a building frame being erected can work safely, thanks to the Sinco Beam Safe, shown here. The cylinder in the cable contains a shock absorber that cushions impact.

chions were set at a height of 3 ft to provide for a waist-high anchorage point for the worker. The system was patented and is currently being distributed by The Sinco Group (East Hampton, Connecticut). Several other contractors came up with variations of the LPR design, with the major differences being in the attachment method and the physical shape and makeup of the vertical stanchion. Other contractors devised ways to attach horizontal lifelines to the structural steel itself, both overhead and at the level of the working surface.

The steel erectors and their workers started realizing major dividends in the form of lives saved. More than a dozen Colorado ironworkers have fallen from the iron since June 1992, all without injury—a win/win situation for the ironworkers, steel erectors, OSHA, and the workers' compensation (WC) insurance carriers.

WC insurance rates in Colorado for steel erection have actually decreased since 1992, from somewhere around $137 of premium for every $100 of payroll to less than $95. While that figure is still high, it shows a dramatic improvement. The ASA Steel Erectors Committee formed a workers' comp dividend group so that they could realize even greater savings for their safety efforts. The loss ratio, which is the total amount of dollars paid out for claims against the total dollars paid in for premium, is currently running at less than 3 percent—an extraordinary achievement for any industry.

The positive relationship that has been developed between the steel erectors and OSHA has continued to strengthen. This relationship is seen as a working tool for both OSHA and the erectors. Region VIII representatives have continued to visit job sites on several occasions to observe the new systems being developed by the erectors. In fact, Assistant Secretary of Labor Joe Dear paid a visit to the Coors Field baseball stadium project in downtown Denver on June 15, 1994. During the tour of the project, Mr. Dear commented that what he was seeing on this job was exactly what contractors across the rest of the nation were claiming could not be done.

After two years of working with various fall-protection systems, we have concluded that the 6-ft fall-protection requirement cannot be met with the technology we have in the field today. Of the 13 recorded falls to date, the workers have fallen from 11 ft to 13 ft before being stopped by the safety device, even though the worker begins to decelerate rapidly after 6 to 8 ft, when the system begins to engage. In a letter to Bart Chadwick dated February 23, 1994, I informed Mr. Chadwick that we could not meet the 6-ft requirement and that the group was adopting an alternative level of 15 ft. The group could not afford the liability of claiming that we would provide complete protection at 6 ft when we really could not.

Shortly after that letter was sent to Mr. Chadwick, another letter was distributed by OSHA headquarters in Washington to all regional administrators around the country, directing them to enforce the old Subpart R regulations, which required protection from falls at heights of 25 and 30 ft, depending on the structure. This interim policy will be enforced until the new Subpart R is promulgated in 1995. The steel erectors have decided to continue with the 15-ft policy we have developed internally, because we know that it saves lives and is achievable.

The people at Region VIII have continued to show support for our cause by providing technical assistance on matters of safety and health. Compliance questions are sent to Region VIII for clarification on a routine basis. The interpretation letters that we receive back from OSHA are distributed to all members of the group for use as training updates and for future reference.

A great deal of the information gained over the last two years in Colorado has been incorporated into the Negotiated Rulemaking Committee that was formed by OSHA to write the new steel erection standards. The committee is made up of representatives from OSHA, steel erection contractors, union leaders, and safety professionals. Several people from the Colorado area are participating.

Colorado Steel Erector Sets an Example with a Tough Turnaround

Three years ago the 100-person, $10 million steel erector LPR Construction (Loveland, Colo.) faced an unpleasant choice: radically reform its safety practices, or fold up. According to contractor safety consultant Steve Miller of Miller Safety (Thornton, Colo.), LPR's safety record was almost unbelievably bad: a lost-workday incident rate of 38, compared with the national average of 14.

Miller told the American Subcontractors Association (ASA) convention in Las Vegas how LPR slashed that workday incident rate to 3—one-fifth the national average and one-tenth LPR's former rate.

Miller, then LPR's safety director, sparked the reforms by instituting these five steps:

- A formal written safety policy;
- Formal training for all, "but it's most important for management";
- Drug testing;
- Formal disciplinary action for those who fail to follow the policy; and
- Incentives for safe work (this includes quarterly gifts and recognition at safety meetings).

As it happens, LPR was not alone. Colorado steel erectors were in the top five among the 50 states in Workers' comp (WC) injury records—their base WC insurance premium was $130 for every $100 in base pay.

Now, however, in a dramatic turnaround, 30 Colorado steel erectors, representing the majority of the state's capacity, are working with OSHA Region VIII to become a national model of OSHA–contractor cooperation. Among the steps they're taking:

- All 30 contractors have adopted the same safety policy, patterned on that of LPR Construction.
- In June 1992, Region VIII OSHA adopted 100% fall protection (anyone working more than 6 feet above the ground where there's no wall or railing protection must wear a safety harness or tie-off).
- All 30 contractors' workers have been safety-trained by Miller, who left LPR to start Miller Safety.
- The 30 have formed a WC insurance buying group that was, at press time, awaiting receipt of three bids from agents competing for their WC business.

The Colorado experience is one model being studied by a committee convened to draft a national safety standard for steel erection. Miller is an ASA representative on the 15-person committee. The OSHA contact is Jerry Reidy, director, Office of Standards and Interpretation (202-219-7207).

(Reprinted with permission from *Contractor's Business Management Report,* July 1993, the Institute of Management and Administration, New York.)

Since 1992, I have been lucky enough to work with two other contractor trade associations, in programs fashioned after the steel erectors group described in this chapter. They are the Association of Wall and Ceiling Industry, Colorado Chapter, and the Colorado Glaziers Association. Region VIII OSHA continues to work with us in pursuing the shared goal of providing a safe workplace for all employees in a practical manner while still maintaining a high level of production and profitability.

For more information on this subject contact:

Steve Miller
President, Miller Safety Consulting, Inc.
Denver, Colorado
(303) 280-9402

Roberta Bourn
Executive Director, American Subcontractors Association
Denver, Colorado
(303) 455-7827

Bart Chadwick
Regional Administrator, Region VIII OSHA
Denver, Colorado
(303) 391-5858

Ed Kassak
Assistant Regional Administrator, Region VIII
Denver, Colorado
(303) 391-5858

Dave Herstedt
Safety Specialist, Region VIII OSHA
Denver, Colorado
(303) 391-5858

Rich Forsberg
J. R. Misken, Inc. (Work. Comp.)
Denver, Colorado
(303) 779-5969

Chapter

12

How One Contractor Partners with Its Employees

Nick Bouler

Nick Bouler, counsel and manager of construction personnel, safety and training for contractor BE&K (Birmingham, Alabama) gives two illustrations showing why his firm is considered a leader in partnering with its employees.

BE&K encourages workers to report near misses—where there was no recordable injury. This is effective for two reasons: (1) It indicates that the company is aggressive about safety, wanting to correct problems before they cause accidents; (2) it gives the program credibility, acknowledging that BE&K needs employees to let them know near-miss situations have occurred. BE&K has an extraordinarily good safety record, and this is one reason.

BE&K is also a leader in partnering with workers on their "financial safety," Bouler writes. Selected employees are invited into the Gold Key program, which guarantees them 2000 hours of work a year; if they do not get that many, the company pays them for the remaining hours up to 2000. And their insurance is paid while they are between jobs.

Why Partner with Employees?

Few businesses are more military in management style than contractors. The superintendent is the "general" of the project; the foreman serves as "platoon leader." Roles are clear. Authority and decision making come down to the "troops" along the chain of command. Challenges to authority are unwelcome and seldom tolerated. In this business, the very idea of "partnering with employees" seems almost un-American. But the goal of partnering with employees is the same as in every other form of partnering—to make the company the most efficient, innovative, and profitable that it can be.

It's the workforce, stupid

President Clinton's campaign offices featured large signs reading, "It's the economy, stupid." The idea was to keep the team focused on the single issue most likely to bring success. I borrowed the idea for a sign in my office that says, "It's the workforce, stupid." That sign helps me stay focused on the single issue most likely to bring success to companies in our business.

More and more of our time as managers is spent on the question: "Where will we find people who have the right skills and are motivated to be productive?" This chapter will look at why this has happened, and suggest ways of thinking that may help answer the question.

In the old days, there was much less of a problem. Superintendents dictated how they wanted things done. "My way or the highway" was a way of life. If the journeyman didn't like it, he could just collect his money and leave. Replacements would be lined up at the gate.

The rewards are not the same

Now, those replacements are hard to find. The temptation is to blame the workforce. How often have you heard (or said) "these kids today don't want to work" or "they don't have the maturity (or initiative, or brains, or discipline) we had in my day." But the causes of change are much more complicated than that.

As long ago as 1981, a skilled craftworker in industrial construction would routinely make $11.50 an hour. That sounds like a fair wage today, in many sectors of the industry. Back then it was real money. A journey-level worker could move around to follow the work and still clear enough money to support a family back home. The wage was especially good compared to the alternatives available for people without a high school education.

Now consider 1995. The minimum wage has risen to $4.25, an increase of 55 percent since 1981. McDonalds pays $5.00 to $6.00 an hour in most of the country, for the most basic entry-level skills, and offers working conditions that are luxurious compared to construction. The Consumer Price Index computed by the government has gone up 62 percent since 1981.

And how has our journey craftsperson fared? He or she can get $14.00 on many large jobs, an increase of only 22 percent *after 15 years*. And many are not doing that well.

The workforce is a resource

While economic conditions have made it hard to make a living in the field, the shortage of skilled people has, ironically, made us see how crucial the front-line employee is to the construction industry.

In many businesses, the product being sold is the knowledge and experience of senior managers. A law firm is like this. In other businesses, the product is a service provided by a well-designed, expensively maintained system. A bank is one example. In these two types of companies the front-line employees are important, but their competence may not be essential to producing a quality product. Mistakes made by front-line employees (tellers, clerks, secretaries) in these businesses are easily caught and corrected. The essential product a customer is paying for is affected, but not controlled, by these employees.

The craftsworker is the product

Compare those situations with construction workers. If a construction company sells a client a boiler repair job to be done during a six-day outage, many skills must be in place. New tubes have to be ordered, received, stored. There are drawings to be analyzed, invoices to be processed, and many questions to be answered. And when all this has been done, nothing has yet happened that a client will pay for.

Everything up to this point is just rehearsal. Showtime is about to begin. If you don't have a pretty good number of people who can lay a t.i.g. root with a hot pass and cap it with a 308 stainless stick, then you are not really in business. Clients pay for a lot of material, but they will only pay for it when it is installed. This is what separates construction from grocery stores and banks and insurance companies. In those businesses the front-line employee only affects the product. In construction, in many ways, the front-line employee *is* the product.

The craftsworker is in control

Naturally, that idea won't be popular with the "old school" field supervisor. He is used to thinking of his workforce as easily replaced cogs. Here is an another idea which, obvious as it is, that type of manager will like even less: The craftsperson is in control.

Most of the money in any project is in equipment and materials. As we know, the customer won't pay for any of it until it is installed in a way that provides a functioning system. The control of all that investment is in the hands of field employees. The best control system in the world will not save that conduit or structural steel or concrete or process piping if poorly trained or poorly motivated employees install it upside down, or don't cure it properly, or cut it off too short, or ruin it trying to make a bend.

The best cost reporting system in the universe will only report the costs those front-line employees have incurred. It won't keep them from wasting or ruining materials and supplies. Only the employees themselves can control costs. Can they be motivated to want to?

The craft workforce is the essential resource

The answer to that last question depends in part on your view of your employees. In the past, some construction managers had a narrow, short-sighted view. They understood that there is only so much steel and concrete and cable that can go into the job. If poor planning or bad techniques caused waste or damage to the materials, replacements had to be ordered and profits became a mirage.

These same managers thought labor was somehow different. They believed that if the workforce was alienated or unmotivated or humiliated to the point where labor was no longer effective, substitutes could be hired at the gate in virtually unlimited numbers, at essentially no additional cost.

If those "good old days" ever existed, they are certainly long gone now. Today we must view field craftsworkers as *the* essential construction resource, the one we dare not misuse. It is not overly dramatic to suggest that the construction companies that prosper in the year 2000 will be those which have learned to train, encourage, and motivate their field workforce.

Why partner with employees? Because they control the company's costs, quality, schedule, and, yes, future. These employees will give a huge competitive advantage to companies that treat them well.

The Theory of Partnering with Employees

Partnering is a way to make your company the "employer of choice" in your area. If skilled and motivated employees would rather work for your company than for the one down the street, you have achieved your goal.

How does partnering, as we will define it, help bring about this happy result? The key is *identification*. In the most successful organizations, employees identify with the company and its goals. Employees are made to feel that company goals will help them to achieve personal goals. Employees internalize the corporate goal because they trust that the corporation and the employee are *going in the same direction*.

Successful companies seem to be able to do the following critical things. If you are competing with these companies for people, these are things you should be doing, too:

1. Give the employee a sense of his or her individual worth to the company. Make clear the importance of each job to the overall goal. Most employees do not think managers realize what the employees do, and the difficult conditions under which they do it. Let your employees know that you do recognize their contribution. I like to tell our field people (because it's true) that no client ever hired BE&K because they wanted Nick Bouler to write a memo for them. My job is to help make the skilled craft employee more productive. The field people appreciate knowing that I know what BE&K's priorities are.

2. Make employees feel that they are key to achieving company goals. This has two parts. First, make it clear that achievement of company goals benefits the field employees. For example, if the employees really believe that the next job will be offered first to people with company experience, they will be a lot more interested in helping you get that next job. We have had a policy of giving hiring preference to former employees virtually since our company started. It has been applied consistently for so long that the employees have faith in it. They support our goals for excellent safety, quality, and productivity because they want us to get that next job.

Second, employees must be made to understand that their work is what the goal is all about. The best safety procedures in the world will be useless if employees do not put them into practice. The best training program will be wasted if employees do not apply what they've learned.

The drawback is that once you have sold your employees on this concept, they will be ruthless toward weaknesses in your system. If materials are not there in time to meet the goal, or if tools or equipment are shoddy or poorly repaired, you will hear about it. Of course, you want that: If employees don't complain about these things, it is because they don't care whether the goal is met. Get their (very justified) pride involved and you will see wonders.

3. Develop a shared sense of achievement as company goals help employees achieve personal goals. A truly successful organization consists of employees working together on common goals, but the willingness to support those goals will arise from each employee being able to achieve his or her private goals with the support of the organization. The goal may be as simple as providing for a family, or it may be to get a promotion to foreman, or to save enough money to go back to school.

Whatever the goal, to the extent the company supports it, the employee will in turn identify the company's goals with his or her own. As we have discussed, this identification is the key to high-performance, high-achievement workplaces. Even the smallest company can have policies the employees recognize as being designed to support them. For example a policy to give hiring preferences to former employees, a policy to give flexible leave for family emergencies, a policy to promote from within. All of these indicate a commitment to be supportive of those who make the company successful.

A Little History

To give some concrete examples of how partnering with employees works, I will talk about activities of my company. BE&K is unique among industrial construction companies in having gone from having a single employee (other than the three founders) in 1972, to being referred to in a recent *Engineering News-Record* article as the country's "most prominent" merit shop contractor. BE&K has also gained national recognition for its employee-centered approach to business.

Because the founders of the firm had significant experience and exposure in the industry, even the first contracts were for very large projects and, in the early years, the question "how will you obtain a workforce?" was a large concern of potential clients. Our chairman, Ted Kennedy, has lasting memories of the days when the very continuation of the company depended on getting craftsworkers to travel to a project and do quality work for a (then) unknown company. In those early years, recruiting those craftsworkers and their supervisors was his job, and he has never forgotten the men and women who accepted the challenge to help make this upstart company into an industry leader.

Many of the people who answered that challenge 20 years ago are still with us, as project managers, project superintendents, and department heads. A large number of our senior field managers worked with Ted when he ran projects in the field. As big as the company is, most of our field supervisors are known to Ted and Mike Goodrich, our president, by name. So when we make personnel policy at BE&K, we have a direct understanding of the people who will be affected.

As a result, no one works for BE&K in any senior management function for very long without being reminded by Ted or Mike that "our assets walk out the gate every night." If we have not provided a job they want to come back to tomorrow, we will be out of business.

In addition, BE&K has made its reputation as a merit shop company. We are not signatories to any bargaining agreements. We look for the best available employees and subcontractors without regard to union affiliation or any other extraneous factor.

This position, which represents 80 percent of today's construction industry, was unusual when the company started. This provided even more reason to be employee centered. We have always felt a duty to prove that our employees do not "lose out" on any needed benefit because of our merit shop philosophy. It's fine to tell employees that they don't need an outside organization to represent their concerns to the company, but when we do that it imposes a special duty to make sure it is true.

We must be doing something right. BE&K is the only engineering-construction company included in *The 100 Best Companies to Work for in America*, by Levering and Moskowitz (Doubleday, New York, 1993). For five consecutive years we have also been named as one of the "100 Best Companies for Working Mothers," compiled by *Working Mother* magazine. BE&K was the subject of the cover story of *Engineering News-Record* for November 22, 1993. The company's concern for its employees was a major focus of that story, which was headlined, "We take care of our own."

Putting Partnering into Practice

There are at least five key areas to explore in putting partnering into practice. We must recognize the *dignity* of the employees, assure their *safety*, make arrangements for *security* for them and their families, keep them *informed*, and provide opportunities for their *involvement* in the company and community.

Employee dignity

It would be nice to think that all supervisors recognize the dignity of every employee. Unfortunately, my experience says otherwise. In addition to the military management style, which tends to demean the front-line "troops" as a way of maintaining command, the construction industry carries over from society a burden of frequent discrimination and harassment of females and minorities.

These issues must be addressed up front. Nothing else that is said about partnering has the least chance of working if employees do not feel that they are respected. Dignity is the foundation that gives credibility to everything else we may try to do.

There is an easy way to determine if you have a problem with this issue. If you think, "Oh, we don't have to worry about that, we'll just comply with the EEO [Equal Employment Opportunity] laws," you are in trouble.

Federal and state employment laws are important, of course, but they represent a bare minimum. We are looking for a much higher standard, one that comes from *you* as a senior manager or owner. It must be clear to your supervisors that *your* standards must be met, not merely those of the government.

Don't let supervisors "manage" employees by cursing and threatening. Why not? Because that's not how we deal with people we respect. And your employees know it.

Recognize that this issue applies to all of your employees. The problem with the government's rules is that, in response, we as managers focus on groups and overlook how often *all* our field people are treated poorly. You may find that many of your managers have decided that the way to deal with EEO rules is to treat everybody equally—equally badly. That attitude, using curses, threats, and shouting to "motivate," will probably keep you out of court, but it drives safety and productivity into the ground.

Your first concern should be to see that your employees get decent (by which I mean "respectful") treatment. The legal issues will then take care of themselves. If your supervisors respect your female employees, they won't ask them for sex in return for raises and promotions. They won't give undeserved promotions to girlfriends. They will talk about the work a woman does instead of speculating on her personal life.

If your minority employees are respected, you won't find all of them working in the civil crafts, and they won't all be working for white supervisors. You won't have people referred to by racial epithets behind their backs. You won't hear "jokes" about intelligence or sexual prowess, which are only funny to people who are usually deficient in both.

Two big points need to be stressed about dignity.

1. Dignity begins with you. Even more than other issues, maintaining dignity is the responsibility of the owners and senior managers. Don't permit racial, ethnic, or sexual comments in your presence. Don't allow wall calendars with sexist pictures in your building, and certainly not in your office. When you begin to make an issue of employee dignity, the supervisors will look to your conduct to see if you really mean it. Raise the level of respect shown to every employee. Dignity is not an EEO issue; it's a human issue.

2. Avoid tokenism. It may seem contradictory, but it is essential for your supervisors to understand that being employee-centered does *not* mean being soft on people. All it means is making tough standards apply fairly to everyone.

In a competitive market, no company can afford to lower it's standards. You must set the bar as high as you can to find the best people to keep the company competitive. What your supervisors must understand is that you will miss a lot of the best people if we just keep hiring folks who look like us.

Current wage rates, which are disappointing to many of our traditional employees, often look pretty good to minorities and females whose opportunities, even now, may be much more limited. At BE&K, we find that employees in these groups can be our most productive and motivated when given the opportunity to show what they can do in even a reasonably fair setting.

Contrary to what many of your supervisors may think, these people are not out there with chips on their shoulders, trying to find ammunition for a lawsuit. Our experience has been that the typical complaint of a minority or female (and we probably don't get as many complaints as we should) is not about harassment at all. The real problem, they tell us, is that they are not taken seriously as craftspersons. They want the pay and promotions that go with their skills, and assignments that give them a chance to show what they can do.

The special case of safety

A sincere emphasis on the safety of field employees is one of the best opportunities most companies have to build good partnering relationships. Safety programs, however, are a little like EEO programs: They have built-in credibility problems because the employees know they are required by law.

One goal of an employee-focused company is to convince employees that the safety effort comes from a genuine concern for their welfare. This concern will be credible if the company shows that it recognizes the field work is what the client is paying for. It also helps to educate employees that the company's safety activities go well beyond what the law requires.

Such a safety program would include the following elements:

High standards, consistently enforced. Find areas that are especially important to your work and go beyond what OSHA requires. BE&K (and a number of other contractors) have gone beyond what OSHA requires in the area of fall protection by requiring the use of full-body harnesses rather than simple safety belts. This equipment was chosen because we do a lot of steel erection and so a number of our employees are exposed to fall hazards. The harnesses spread out the force and so are much less likely to cause injury to an employee who falls than do the belts.

The harnesses do cost more, but the employees know that we are not required to do it, so we gain something in their eyes. The surest sign of commitment in the business environment is the willingness to spend money.

Equally important is consistency in applying your standards. Safety rules have to apply to everyone, but if employees are encouraged to partner with you, they will be enforced by everyone. When the project superintendent comes out of the trailer and is stopped by a laborer for not wearing safety glasses, you can begin to feel that your program is really working.

Special emphasis programs. At BE&K we have several programs designed to help get the spirit of partnering in safety out to the front-line employee. On some jobs, every hardhat sports a red sticker with large white letters reading "It's Okay." Employees wearing that sticker have signed a "contract" that it's okay for any other employee to stop him or her to correct an unsafe work habit, and has agreed not to get upset or defensive.

On other projects, large round stickers with a red cross inside a circle read "I have the authority—to make the work safe." (See also Fig. 12.1.) Employees wearing these stickers have been trained to understand that this includes the right to suggest that work should be stopped if there is a question that unsafe methods or conditions may be present. This is a big cultural change from the "military" style, and there have been concerns that it might be misused. But we have not had a problem with that happening so far, and I believe the potential gain from having people speak up *before* an injury or damage occurs will far outweigh any occasional misuse.

Site safety committees. Every BE&K project has a site craft safety group, including at least one person from each craft, which meets with the senior safety person weekly. This group conducts inspections and serves as a clearing house for suggested improvements in the safety process or items needing correction out in the field. Typically the committee is composed of journey-level workers and helpers only. No supervisors are part of this group, to encourage free expression by all members. Any deficiencies are written up and given to site management for immediate correction.

Reporting of near misses. Many construction organizations are seeing improvement in the safety process as a result of rethinking which events should be getting management attention. Too often in the past, we did not really consider something to be a problem until the condition or work practice resulted in a recordable injury.

Now companies are beginning to realize that many serious accidents result from conditions or practices that had been in existence

Figure 12.1 With this medallion, BE&K reinforces the message that workers are to put safety first, and that workers have the right to suggest that work be stopped if there is any question of unsafe methods or conditions.

long before the injury or damage, but that had not been recognized. Having the employees feel that they are partners in the safety effort can dramatically improve this situation.

Asking employees to report "near misses" is an effective tool to draw them into the safety process, for two reasons. One, it indicates that the company is aggressive about safety, wanting to correct problems before they cause accidents. Two, it gives the program credibility because it acknowledges that we need the employees to let us know that these situations have occurred.

When we started this program, our supervisors were leery. They were concerned that the home office was just coming up with another statistic to use against them. They said, "How can you tell what a near miss is?" I told them, "A near miss is when you hear what happened and you wipe your forehead with the back of your hand and say 'Whew!'"

Gradually they have seen that near-miss investigations are the best way to learn where the process may have a glitch, so it can be fixed before injury or damage occurs. And there is a real bonus when front-line employees are involved in the investigation. When they see time and effort put into finding a problem *before* a serious injury occurs, they know that your company's safety commitment is genuine.

Celebration of successes. Part of any successful organization is creating a team that people are proud to join. Nothing develops team spirit better than sharing accomplishments. Find reasons to hold little celebrations. If your group works 100,000 hours without anybody missing work because of an occupational injury, that may be reason enough to provide biscuits in the morning, or to pass out candy bars at the gate at the end of the shift.

These examples may seem trivial when a big company may easily spend $25,000 on a picnic for a 1000-employee project that works 2 million safe hours. But the employee who gets a candy bar from the boss at the gate may feel much more recognized and rewarded than one attending the picnic, if no senior people are there and the preparation and emphasis have been lacking.

If you are proud of your people and you communicate that pride in a direct and sincere way, whatever you do will be received gratefully. Try something!

Family-centered benefits. The whole issue of benefits is a sensitive one for the smaller company. Obviously the company has to watch costs: This is not optional if the goal is to win contracts. Again, however, it may be possible to reap benefits with employees by (1) communicating and (2) doing what you can afford.

BE&K uses the term "family friendly" to describe a benefit program that we have tried to make very comprehensive. This term recognizes a universal truth: Most of your employees are working to provide for their families.

A successful benefits program recognizes this truth as a value that employees have in common. A sure way to win the loyalty of an employee is to help her have insurance when her child gets sick, or have time off to care for an emergency, or to have security of employment.

BE&K provides life and health insurance to all hourly employees after 90 days. The employee's coverage is provided at no cost, and BE&K pays 50 percent of the family premium portion. The level of benefits and deductibles is the same as for the salaried employee plan.

Even before the federal government enacted the Family and Medical Leave Act, BE&K provided flexible leave to deal with family emergencies.

One of our best-known benefits has been the provisions we make to assist our employees with child care. "BEKare" is a nationally recognized, high-quality child care center that was based in modular structures at one of our largest job sites. It has now been formed as a permanent structure serving the home office in Birmingham.

In addition, we provide partial subsidies for child care to employees at our job sites, and fully subsidize the cost of after-hours care when parents are required to work after regular hours.

For our hourly employees there is often no greater benefit than job security. BE&K has made a start on this issue by establishing a "Gold Key" program (see Fig. 12.2). Selected employees are chosen (based on qualifications and loyalty to BE&K) to become Gold Key employees. For this group we *guarantee* that we will offer 2000 hours of work per year. If we do not offer at least that much work, we pay the employee for the hours we did not offer, up to 2000.

Figure 12.2 Under BE&K's "Gold Key" program, selected employees, chosen based on qualifications and loyalty, are guaranteed 2000 hours of work per year, or are paid for hours under 2000 for which there is no work. Their insurance is also maintained during nonwork periods.

Gold Key employees also have their insurance maintained while they are between jobs. We continue the company's contribution to the premium even though the employee is not on the payroll.

To continue the emphasis on families, we make available two scholarships per year for the children of Gold Key employees. An additional two scholarships are available for children of salaried employees.

Smaller companies will have different programs, but these examples may give you some ideas. Let employees know you are looking at alternatives that will let the company remain competitive. Look to your industry associations for help. Both ABC and AGC have been leaders in helping smaller companies obtain group rates for innovative plans.

The informed employee

One part of the theory of partnering is that employees will help the company reach its goals if they know what the goals are and if they understand how the goals relate to success. The old military style of saying "charge that hill" without explaining the objective or asking for buy-in and participation is not likely to work today.

Informational meetings. Periodically we hold meetings with sizable groups of employees and a senior manager telling where the company is headed, what the business environment looks like for the next six months, and taking questions. These are not always sunny affairs, and we do open ourselves up to some awkward questions, but overwhelmingly the employees view these occasions as one of the ways BE&K is "not like other companies." This is a concept that can be implemented by any organization.

There is some small cost involved in the employees' time, and our meetings frequently include a light lunch, but if attendance is voluntary you could hold a meeting while employees eat their lunches. Perhaps you could use a half-hour (voluntary) of their time and half an hour on the company.

All of us want to feel "in on" the things that are important to us. Employees who see you discussing real issues in a candid way will be better members of the team.

Teaching the business. A somewhat related activity that has been extremely popular in our home office is a training program to show employees in an informal way what the company does. We are big enough that we have bookkeepers, buyers, draftsmen, secretaries, and service employees who have never been to a job site and who do not know what a construction company does.

In our EPC training series, a group of employees meets for one hour each week for 12 weeks. At each meeting, a department head gives a presentation about which employees are in that department and what, exactly, that department does. Each presenter describes his or her department's work in terms of what was done on an actual project. In effect, the group can visualize the project going up week by week, as the work is described for them at each session—from the designers, estimators, purchasing agents, and expediters, right through to the instrument techs and start-up engineers.

Special project groups. One interesting thing you will learn when you start to bring your employees into the information flow is that they will begin to get excited about where the company is going, and they will want to do more to help it get there. Many problems in any organization are not solved because the front-line person who deals with the bottleneck every day is used to working around it, cursing as he or she does so. If those employees begin to believe you care about what they know, you will probably find that they have ideas about how to clear up those bottlenecks. Solutions will come from the people who know the most about the problem.

Our secretaries in the home office let us know about a problem with our filing system. Each department handled its records differently. We had no standardization about how files were labeled, which were stored where, or when they could be destroyed to make more space. The secretaries formed a study group to recommend a comprehensive system to deal with this issue.

Here's one tip: Make use of *all* your people's talents. I think it is 3M that has that great slogan, "Every employee comes with a mind, at no additional cost."

Teaching the crafts. It is easy to overlook the importance to employees of being able to learn a new skill. In these times, when many feel that public education tends to neglect the children who don't plan to go to college, it is no small thing when a young person can come to your project and learn a trade.

BE&K long ago recognized the importance of craft training to the company's future. We offer classes after work on virtually all job sites, and we like to say that on any night of the week hundreds of people all over the country are in BE&K classroom trailers learning to read prints, or bend conduit, or supervise crews.

Since 1989 we have had more than 40,000 completions of at least one training course. Many employees taking these courses moved up more than one level (and pay grade) as a result of this training. And we find that they think of themselves as BE&K people, because this is where they learned the business.

Like the other ideas in this chapter, this one can be applied in both large and small companies. While training is very expensive, there are ways to find opportunities for your people. ABC offers the "Wheels of Learning" courses to its members, and similar training is available through AGC. Smaller contractors often find that access to these programs alone is well worth the cost of membership.

Providing training is a true "win/win" situation. The employee learns a way to make a living at pay rates well above what many entry-level employees are able to find in this new "service economy." And your company gets a motivated and grateful employee who has a lot of reasons to identify with your goals.

The involved employee

Providing opportunities for employees to be involved in community activities has been a very successful partnering technique for BE&K.

Adopt-a-School. We were among the first construction companies to use "Adopt-a-School" programs. The project site forms an alliance with a local school. In many cases, where the construction job is in a rural area, the choice of school is easy. At other times there may be a number of candidates. Generally, we try to pick the one that needs the most help.

Our employees work on improvement projects for the school (see Fig. 12.3). The company donates materials, and employees donate their time and skills. Our people have built band rooms, shower facilities, installed new windows, replaced antiquated wiring systems, and generally done about every kind of renovation project imaginable.

Figure 12.3 Under its "Adopt-a-School" program, BE&K helps a school in each community where it has an ongoing job. The company donates materials, and employees donate their time and skills.

Since many of our skilled employees and supervisors travel away from home to work with us, these projects provide an opportunity for our employees to feel connected to their temporary communities and to leave a lasting contribution to the towns that welcome them.

Community service grants. We have developed other programs that are oriented toward employee involvement in the community in tangible ways. One of these is our grant program. BE&K saw that a number of large companies would match donations made by employees to charities and service groups. This seemed like a good idea, but it also seemed there might be a way to improve it. In our plan, BE&K makes donations, based on an employee's application for a grant, to organizations the employee actively and regularly participates in.

All employees have an equal chance to benefit their favorite organizations. With most "matching" programs, an executive might be able to donate a $1000 gift (to be matched), while another employee might be hard-pressed to donate $50.00. And you don't spend the corporation's profit sharing on groups that may be of only minor interest to most employees.

Again, we found a way for corporate goals to support the goals of the individual employee.

Conclusion

By now it should be apparent that partnering is very different from the old style of ordering employees. The goal is to have shared values and goals that result in gains for both sides.

Partnering with employees offers a way to take your company beyond the limitations of military-style management, to become an organization which attracts the skilled people that are a key to success.

Do what you can to recognize their dignity, provide for their safety, help them achieve some security, keep them informed, and give opportunities for their involvement, and your employees will repay your efforts with a lasting competitive advantage.

I would be happy to respond to written comments or questions about the content of this chapter. Please contact me at:

BE&K Construction Company
P.O. Box 12606
Birmingham, AL 35202
FAX (205) 972-6135

There are many excellent sources of additional information about issues involved in relating productively to the workforce. Among many others, the following are recommended:

Robert Townsend, *Up the Organization.* Fawcett, New York, 1983.

Tom Peters, *Thriving on Chaos, Handbook for a Management Revolution.* Knopf, New York, 1987.

National Center on Education and the Economy, *America's Choice: High Skills or Low Wages!* The Report of The Commission on the Skills of the American Workforce, Rochester, NY, 1990.

Robert W. Dorsey, *The Acquisition of Skills and Traits Among Construction Personnel,* Source Document 54, Construction Industry Institute, Austin, TX, 1990.

Update: Minority & Women Workers in Construction

If your construction company is led and managed by white males—as 99% are—you probably see and work with relatively few minority and women co-workers/peers. But the world is changing, contractors were told at a recent Associated Builders and Contractors' (ABC) convention in Miami Beach. The number of contractors that can survive by limiting their hiring to white males is shrinking rapidly. *The Wall Street Journal* recently spotlighted Phoenix housing contractors' desperate search for skilled workers. And at the ABC meeting, president Steve Westra of Westra Construction (Waupun, Wis.) said he can no longer get carpenters using customary recruiting methods.

The "Workforce 2000" report of the Hudson Institute predicts that in six years, only 20% of work force entrants will be U.S.-born white males.

Hiring minorities, women, and immigrants now can only strengthen a contractor's work force. Nick Bouler, counsel and manager of construction personnel, safety, and training for BE&K (Birmingham, Ala.) said, "I want ABC members to know we're not talking about lowering hiring standards [to the contrary—because women and minorities] tend to be better motivated.

"Male craftsmen seem to feel they can always find another job." This is not so with the other groups. "Look at small towns [where most of BE&K's paper mill jobs are located]. The job alternatives that women and minorities have in many cases are not very attractive." (For more on BE&K's hiring policies, see sidebar below.)

Paula Clements, executive director of the National Association of Women in Construction (Ft. Worth), told ABC that 800 of the association's 7,000 members have started their own construction firms, which have revenues up to $20 million. Many struck out on their own because of a lack of opportunity in the companies for which they'd worked.

Once hired, treat women and minority workers no differently from the white male workers.

Said Bouler, "At BE&K we care about productivity. Make it clear to your supervisors—the ones who will be most concerned—that you have no intention of compromising quality of work."

Bouler said when he interviewed women workers at a Georgia jobsite, their biggest complaint was, "They won't let me be a journeyman. I'm better than that guy, and he's getting $14.50 an hour while I get $11.75."

Added Bouler, "You don't have people out there wanting to sue you, you have people wanting to be treated fairly."

Try to retain the women and minorities you do hire. Once your company becomes known as one that hires, retains, promotes, and develops minorities, they'll want to work for you. Recruiting becomes no problem.

Hiring is only the beginning. Although contractors are overcoming their resistance to hiring qualified minorities and women, many are still not promoting them.

"We found that barriers to promotion do exist," said Clements, a veteran of the Glass Ceiling Commission appointed by former president George Bush. For example, a company will go outside for an executive-development candidate, rather than look at a qualified minority employee within the firm.

One ABC-member contractor that admits to having this problem is Nova Group (Napa, Calif.). Carole Bionda, VP and counsel, explained that Nova is aggressive about hiring minorities—they constitute 70% of its field crews. But "field management was all male. We were hiring them, but they weren't working their way up."

So Nova held a two-day retreat facilitated by consultant Marjorie Bradford of MTB Enterprises (Cincinnati).

(Continued)

Recalls Bionda, "You could almost see our superintendents holding back. But as the first day of the retreat wore on, they opened up. Bradford told them,'It's okay to feel more comfortable with people like yourselves.'" She made them understand it was natural to feel threatened.

"We had to empower them to say [to all employees], 'If you are good you'll stay, and if you're not you won't.' One of our problems with females was that superintendents made it too easy for them. The women resented it."

Minority subs. Although the ABC session focused on contractor hiring of minorities and women, hiring a minority subcontractor is an alternative way to accomplish much the same thing. The following provide tips on how to find minority contractors:

- The National Association of Minority Contractors (NAMC, Washington, D.C., 202-347-8259) "does an incredible amount of legwork for me when I go into a new state seeking work," said Nova Group's Bionda. "It has directories of minority contractor specialists—drywall, electrical, and so on."

    NAMC has minority supplier members, too, reported Kirkwood Bolton, president of the minority contractor A.G. Gaston Construction (Birmingham, Ala.).

- State agencies such as DOTs keep lists of minority contractors, Westra said.

- The Small Business Administration (Washington, D.C., and regional offices) has published a directory of female-owned construction companies.

(Finally, *The Best Companies in America for Minorities* tells how to recruit, train, and promote a diverse workforce.)

Do BE&K's Diverse Workforce Policies Represent the Future of Construction?

The large industrial contractor BE&K (Birmingham, Ala.), a leader in hiring minorities and women, may be a workforce-diversity model for the construction company of the future. Nick Bouler, BE&K counsel and manager of construction personnel, safety, and training, told the ABC contractors:

- BE&K, which builds large papermills throughout the southeast, often requires 1,000 people on a job. Being a transient business, construction presents special personnel problems. Usually, 30% of a construction crew travels from job to job. "In our workforce, many spouses and partners travel with their men. Over the years we've hired a significant number of traveling spouses, many of whom in time have become very skilled craftspeople."

- BE&K has won kudos as a desirable place to work. One example of that has to do with its daycare service. Bouler said, "If a worker's child is sick, we'll subsidize a daycare center to keep its doors open" for as many hours as the worker needs.

- The most recent BE&K craft-training class at Plymouth, NC included 17 instrument fitters, only four of whom were white males. Six were white women, two were black women, and five were black men. Before, "If we had blacks on a job, they were laborers or concrete finishers." No longer.

Reprinted from *Contractor's Business Management Report,* with permission of the publisher, Institute of Management & Administration, New York.

(Continued)

When a Family Member Doesn't Work Out in the Family-Owned Construction Company

One of the toughest issues facing the head of a family-owned construction company is how to live with a family member who enters the business and, for whatever reason, is unhappy or ineffective. Steve Westra, president of Westra Construction (Waupun, Wis.), recalls one such case:

"A young friend of mine took over his parents' construction business. Since they both died relatively young, he not only ran it, he saw all his brothers and sisters through school. Several siblings came back to work in the company, but one didn't work out, partly because his wife was unhappy that the president was taking home more than her husband. She forgot that [the president] was working 80-hour weeks. The situation became untenable. After agonizing over what to do, my friend brought his brothers and sisters together and told them, 'You have two choices. Either I buy you out or we each auction off our stock.' It took everyone a while to realize he was serious. Ultimately, they agreed to sell out to their brother."

Working in a family business, concludes Westra, can be the best possible arrangement—or the worst. "When it works, there's no better environment. And when it doesn't, there's nothing worse. Family emotions can strain relations. The business world is littered with casualties."

Reprinted from *Contractor's Business Management Report,* August 1993, with permission of the publisher, Institute of Management & Administration, New York.

Chapter

13

How One Design Firm Partners with Foreign Coworkers

Henry L. Michel

Ask the head of any successful A/E/C organization how the firm was built, and he or she will tell you, in effect, "partnering with our coworkers and clients." That's equally true when the firm has foreign offices. Henry Michel, chairman of the engineering firm Parsons Brinckerhoff (PB; New York City), has a 10-point checklist of steps the firm has used to build the largest foreign design practice of any U.S. firm—600 people overseas, 15 percent of PB's total staff. Among Michel's points are:

- Staff it largely with local professionals (only six of the more than 350 in PB's Hong Kong office are non-natives).
- Encourage and assist your foreign coworkers to develop technically. If you don't share technical strengths through training and staff transfers, you will soon be categorized as a colonialist, simply trying to enrich yourself at the expense of the local community.
- Make all foreign employees eligible for all benefits, including stock purchase. If you are not prepared to give up something so the local staffs (especially the local management team) can feel a sense of belonging, your initial overseas marketing successes will be short-lived.

Much has been written about partnering in the construction industry—between different engineering disciplines, between designers, constructors, subcontractors, suppliers, investors, owners, and so on. Regardless of the form of partnering, one common denominator determines the success or failure of any teaming approach: people.

This book will not result in any improvements in delivery of a finished product if we do not care for our basic inventory: our people. Teamwork is not a manufactured product; it is the result of careful nurturing, training, motivating. An earlier leader at Parsons Brinckerhoff used to say that our inventory goes down the elevator every night. To which we should add that it is our responsibility to ensure that the same inventory wants to go up that elevator the next morning.

So, partnering begins at home, by partnering with your own coworkers. That is easy to do with staff close at hand, at your own headquarters office. In our global environment, however, distance does not always make the heart grow fonder—except maybe for another employer.

In the mid-1970s, when our firm was nearing its 90th year, a case of mid-life crisis or, more likely, of hardening of the partnership arteries had set in. The small number of aging partners who had preserved their elitist collaboration without seriously spreading the wealth saw a problem of ownership transfer which could only be overcome by selling the firm. Fortunately, a small minority of younger partners were able to block the sale. That part turned out to be easy, but then we were faced with the need to turn around a moribund, underfinanced company.

The very first step, and the one which saved us and built the foundation for our present preeminence, was to make partnering with our key employees a top priority. We wanted not just a tiny handful of omniscient, all-powerful partners, but a much broader band of key employee owners. We converted to a corporate form of organization and created a Key Employee Stock Ownership Plan or KESOP—not tax driven but recognition driven. We felt that every professional services company has a group of employees who should never need to be replaced, and we fixed the size of that group rather arbitrarily at 10 to 15 percent of total employment—not the 2 percent or 1-in-50 ratio of the old partnership days. Of course, the other 85 to 90 percent of the staff were important and valued, but if they were to leave, people of at least equal competence would be available as replacements.

We decided that our KESOP had to be structured to capture the collegiality of partnership with the incentives of an entrepreneurship: a partnering with our own staff. The resulting recognition, the sense of belonging and purpose, and the sharing of a "piece of the rock" have resulted in growth of ownership to the point where we now have nearly 500 key employee stock owners or "partners," totaling about 12 percent of our present staff.

As we began to grow, the question arose: "How can we build the internal partnering concept into an international practice?"

International Practice

There are considerable differences between truly international work and performing work at home for a foreign location. To be truly international is to work in a foreign environment with foreign engineers. It takes a different kind of training and a different kind of commitment than the occasional foreign project that is designed in the United States and in which one or two transplanted Americans manage the operation overseas. To go international requires a true partnering approach, with a willingness to share ownership in a foreign land, to share management functions, and, most relevant here, to share technology. There are obvious difficulties, but the rewards are considerable and the potential is tremendous.

Parsons Brinckerhoff has a long history of working in foreign countries. As early as the Great Depression, the firm's leadership sought work in South America. The first contracts were won in Venezuela for a complex of maintenance shops at a mining company's highland iron ore operation, and a water supply system for the city of Caracas. Hydrological projects soon followed in Peru, Ecuador, Argentina, and Colombia. Although most of the designs for those projects were prepared in Parsons Brinckerhoff's New York office, the local staff included South American engineers as well as imported North Americans.

By 1942 the firm had combined its Venezuelan and Colombian subsidiaries into the Venezuelan Corporation of Consulting Engineers, Compania Anonima. There were 70 employees, 50 of whom were Venezuelan and Colombian professionals.

When a change in government in 1945 terminated U.S. and European contracts in Venezuela, our Venezuelan employees started their own consulting engineering offices modeled after Parsons Brinckerhoff. The same happened in Colombia when that country's government terminated foreign contracts in 1946. There, Parsons Brinckerhoff's leading Colombian engineers formed OLAP (Olarte, Ospina, Arias and Payan), now called INGETEC, which has become one of the largest engineering firms in South America and a welcome member of future partnering arrangements.

These countries, which had practically no consulting engineers when the firm began its first South American projects, had learned both technical and organizational skills from working with Parsons Brinckerhoff. Painlessly and perhaps unwittingly, Parsons Brinckerhoff had trained a generation of engineers who were willing and more than capable of managing their own operations. Since those first experiments in South America, Parsons Brinckerhoff has worked all over the world and has consciously developed people-partnering programs to fit a variety of culturally diverse circumstances. In most cases, and I shall describe a few, those programs have contributed significantly to our successes abroad and have led to long-standing cooperative ventures that have benefited both the firm and our foreign counterparts.

North Africa and the Middle East

After South America, where the firm has worked continuously from the 1930s to the present, Egypt was a totally different climate. The long-term British presence there had created a certain managerial sophistication, and Egyptians were already managing their own consulting firms when Parsons Brinckerhoff arrived on the scene in the mid-1970s. However, there was room for improvement. In preparation for the Cairo Ringroad project, Parsons Brinckerhoff's first major contract in Egypt, the firm brought a very senior Egyptian civil engineer to the United States for training. Armed with newly learned technical as well as administrative procedures, he returned nine months later to serve as deputy manager on the project and eventually became operating manager of Parsons Brinckerhoff's Egyptian subsidiary.

Of necessity, and productively, Parsons Brinckerhoff worked closely with the Egyptians on that first project and on subsequent projects. For the Cairo Ringroad, built for the Ministry of Housing and Reconstruction of the Arab Republic of Egypt, the firm joined forces with Sabbour Associates of Egypt and Ward, Ashcroft and Parkman of England. It again joined forces with Sabbour (and also Kaiser Engineers) for master planning, engineering, and design services for port and harbor facilities at Suez City, and later in yet another partnership with architects and urban planners to produce the Master Plan Study for Sadat City.

Throughout those collaborations, which resulted in the founding of an Egyptian company, PB Sabbour, in 1979, the flow of knowledge was far from one-sided. To work successfully in this totally alien environment, Parsons Brinckerhoff needed its Egyptian colleagues to guide it through the maze of cultural differences as well as to provide insights into the complex issues that had to be dealt with on such projects. Cairo's Ringroad was intended to streamline traffic flow but also needed to reflect the city's historical heritage, its present accomplishments, and its future prospects. Those issues obviously could not be mastered overnight and certainly not without considerable assistance from our local partners. To rebuild Suez and plan Sadat City—an entirely new metropolis halfway between Alexandria and Cairo—similarly required more than American technical expertise. To this day we rely on our local partners, who are now working side by side with us in managing the implementation of a modern urban rapid transit system, the Cairo Metro.

From Egypt, Parsons Brinckerhoff went to Morocco. There the firm joined a multinational consortium to design a 602-mile-long extension of the Moroccan national railway system. Parsons Brinckerhoff had purchased half of Central Technology, Inc. (the other half was owned by an Italian group), in 1977, which resulted in the formation of PB CENTEC. In exchange for teaching the Moroccans about railway engineering, Parsons Brinckerhoff fostered a local subsidiary company, an organization of local professionals, eventually called PB CENTEC Maghreb, and we provided them 25 percent ownership.

We were then in a strong position to export services to nearby countries, including the Sultanate of Oman, where the part-Moroccan-owned affiliate conducted a major flood control and water resources project. In Tunisia, those newly acquired technical skills, coupled with linguistic (French and Arabic) and cultural capabilities, were applied to a water reclamation program. However, the workload available to this highly honed, locally managed group was insufficient to maintain our continuing presence there, so we transferred the entire staff to our client, the Moroccan National Railways Company. Now that organization has its own, in-house trained engineering capability—which had been denied them by their previous colonial sponsor—and we have a future client partner.

The general approach in North Africa and the Middle East has been to share American technology in return for obtaining culturally appropriate and already developed local management skills. The situation was not the same in the Far East.

The Far East

The cultural milieu in the Far East was different. There were no potential indigenous managers to bring to the United States for training. The British in Hong Kong, unlike those in Egypt, had kept the decision making and leadership in their own hands while delegating the work to the Chinese. So entrenched was the British system that it was difficult to convince the Chinese that it could be otherwise.

Parsons Brinckerhoff began its entry into the Hong Kong market in 1977. With its newly proven expertise on subway ventilation systems, the firm was able to compete successfully for the design of the Hong Kong subway's environmental control system (SES). Hong Kong's Mass Transit Railway Corporation retained Parsons Brinckerhoff for SES analysis and design for the 7-mile Tsuen Wan Branch Extension in 1977 and for conceptual development of an environmental control system for underground stations for the proposed new Island Line on Victoria Island.

Seeing great potential for more work in Hong Kong and a unique opportunity for growth, Parsons Brinckerhoff set up permanent headquarters—Parsons Brinckerhoff (Asia) Limited (PBA)—in 1979. Under the leadership of an expatriate manager who had learned the business intricacies of Hong Kong, and with the help of a long-time Chinese-American employee of the firm and now PBA's president, we developed an all-Chinese staff for all levels of responsibility. Chinese were for the first time offered a role in management. Rather than importing project managers from the United States, our expatriates trained talented Chinese personnel locally to do the job. By 1982, PBA boasted a staff of 70, which has since grown to more than 350.

The Chinese proved to be tremendously industrious, anxious to learn, and totally capable of handling the increased responsibilities. Local participation engineered a loyalty and pride in the company that benefited both Parsons Brinckerhoff and the Chinese personnel. With access to upper-management positions, the Chinese were able to contribute creatively to the technical and marketing sides of the business, to receive credit and rewards for their contributions, to take pride in the work done, and to become shareholders in the parent company.

Although the technology transfer—and what might be called organizational transfer—began as a one-sided process, it quickly developed into a two-way street. By working side by side with the Chinese, our expatriate Americans learned the local practices and identified another niche market—building services—which has led to numerous contracts in that area. One example is the Hong Kong Convention and Exposition Centre, shown in Fig. 13.1. By entrusting responsibility to Chinese project managers, the firm gained a cadre of effective new leadership with invaluable connections to local clients.

Most important, the firm gained a core staff that was ready to export their services to the other Pacific Rim areas. That group became the basic cadre for opening up new companies in Singapore, Malaysia, and Thailand. Those companies are totally locally staffed and managed by transplanted Asians whom we trained at their home base.

The Singapore office, now a local company but started by a core group of PB Asia-trained Hong Kong engineers, has started to adapt

Figure 13.1 Parsons Brinckerhoff's aggressive partnering with international engineering coworkers is only one part of a multipoint strategy for winning international jobs. The PB project pictured here—the Hong Kong Convention and Exhibition Centre complex—illustrates another PB tactic, superior technical strength. Several years ago, PB noted it had significantly more strength in mechanical/electrical engineering design than did the British design firms which dominated the Hong Kong market. So PB aggressively sought mechanical/electrical assignments, and has won a large number in Southeast Asia.

the best features of a truly indigenous entity. Its more than 100-person workforce has captured a significant portion of the local market and has exported services to Vietnam, Malaysia, and Myanmar. Now a Singapore-trained core group will be the start-up cadre for a new Malaysian company. Closing the loop, Asian-trained engineers/stockholders are now helping staff work at JFK International Airport in New York, the Cairo Metro in Egypt, the Jubilee Line of the London Underground, and so on.

Japan was—as our British friends like to put it—yet another cup of tea. The country has been largely closed to foreign influences, but the Japanese have always been aggressive exporters of goods and services. In the early 1980s a rare partnering opportunity arose. We had completed the World Bank–financed road and transit master plans for the city-state of Singapore some 10 years earlier. President Lee now called for design-build bids from hungry international contractors. The major Japanese builders and suppliers were anxious to participate, but there was a small problem: President Lee had once been incarcerated in a Japanese concentration camp and had less than fond memories of those days. However, he thought highly of American technology.

So we were approached by two Japanese consortia, one for civil works, the other for electromechanical installations, and we started relationships which have continued to this day, with partnerships with Japanese trading companies, contracting groups, and suppliers on projects in Thailand, Australia, the People's Republic of China, Hong Kong, Korea, and Japan itself. One of many joint programs includes utilizing the services of young Japanese construction engineers on our stateside construction management projects, while in exchange our young engineers get an opportunity to work in Japan—another great people partnering approach.

Turkey

In Turkey there was yet a different approach. As part of the assignment for the Bosphorus Tube/Istanbul Metro design, the Turkish Ministry of Public Works and Settlement required "methods by which the Consultant and Government staff could work together on the project." The ministry not only wanted to participate in project management (through a project officer and supporting staff within the ministry), they also wanted a program to train selected employees for later direct participation in the project.

In response, we created a training program that went beyond the usual technology-exchange framework. Qualified Turkish engineers were not just brought to the United States for theoretical training, but were put to work on actual transportation projects—tunnels, metros, highways, and bridges—that required skills related directly to those needed on the Bosphorus Tube/Istanbul Metro project. They were given the opportunity to "learn on the job," not only the technological aspects of state-of-the-art American engineering, but also the procedural aspects—how Americans work, how we think, how we do business. Through that program, 20 Turkish engineers spent one or two years working in our various offices around the country.

There were, of course, prerequisites for participation in such a program. For example, only engineers with a mastery of English were eligible, and good academic training and some practical experience were necessary.

The benefits of the program were incalculable, on both sides. At no cost to the Turkish government, their own engineers have been prepared to take on major technical, project engineering, or management roles in railroad, tunnel, and metro projects and have provided Parsons Brinckerhoff with a competent technical resource for future local projects.

Summary

The success and growth of companies in the services industry is totally dependent on various forms of partnering. The obvious ones, such as partnerships with owners, with builders, with suppliers, with regulators, with crafts unions, etc., have been widely covered in seminars, conferences, and writings. The more subtle aspects relate to the people side of the business—to what degree we succeed to bridge the gaps between "them" and "us." This requires the willingness to share and to give openly—not grudgingly—to provide training and opportunities for applying the new knowledge, to share responsibility and authority, and to share rewards including the ultimate reward—ownership in your enterprise.

If you achieve this, as we have at Parsons Brinckerhoff, then you have mastered the art of partnering.

Parsons Brinckerhoff Builds Its International Practice through Partnering with Its Own Staff

Henry Michel, chair of the engineering firm Parsons Brinckerhoff (PB; New York City) may have more overseas experience than any other U.S. design-firm principal. He has worked overseas for 30 years, including 15 years as a civil engineer; he has run an A/E firm in Rome; and now he heads a firm at which nearly 600 of a total of 4,000 employees work overseas. In Taipei, Taiwan, a joint venture of PB, Bechtel, and Kaiser Engineers is currently designing and building a multi-billion-dollar transit system. In Egypt and Turkey, PB has created companies jointly with strong local A/E firms. And in Thailand, Australia, and Taiwan, it has teamed with non-local firms to pursue local work. Among Michel's suggestions on building an overseas practice:

1. Establish a local company.

2. Staff it largely with local professionals (only six of the more than 350 in PB's Hong Kong office are non-natives).

3. Encourage and assist your foreign workers to develop technically. If you don't share technical strengths through training and staff transfers you will soon be categorized as a colonialist, simply trying to enrich yourself at the expense of the local community. PB recently brought two Chinese coworkers—its Hong Kong office's chief ME and chief EE—to New York City to help design the central refrigeration and steam plant at John F. Kennedy International Airport.

4. Support the office initially with specialist resources from your other offices.

5. Realize the toughest jobs to fill (as in U.S. offices) will be management posts such as project managers.

6. Make all employees fully eligible for all benefits, including stock purchase. The PB philosophy is to treat foreign coworkers as equal partners. Since 1975 PB has restructured ownership to enable about 12% of coworkers at all levels—those considered most valuable—to become stockholders. This benefit applies equally to foreign coworkers. If you are not prepared to give up something so that the local staffs, especially the local management team, can feel a sense of belonging—your initial overseas marketing successes will be short-lived.

(Continued)

7. Make maximum use of staff assigned to the job by your overseas client. Then they can learn new skills to apply in the future on similar projects. PB did this in the '70's and '80's with Atlanta rail-transit client MARTA. The two organizations paired many individual engineers in a trainer–trainee partnership. Now that model has been adopted in Taipei, where PB, Bechtel, and Kaiser Engineers have a huge transit-system design and construction-management contract.

8. Spread the word overseas about how your firm and its local counterparts are working together in new ways. Encourage the partners to co-author papers, give speeches and seminars, and so on. This gives the firm local credibility, which will improve its chances of winning future assignments.

9. Don't overlook your in-house staff who were born in your target nations. They may have interesting contacts there. Identify employees of foreign origin in each country of interest to your firm; make them your mentors on local issues.

10. Make full use of foreign-born engineering graduates of U.S. institutions. Some 60% of U.S. engineering PhDs awarded in recent years are to foreign born. They can form the nucleus of your international practice.

11. Team with other non-local firms that already have a local presence. Partner with them in a joint venture. They will say yes a surprising percentage of the time, especially if you can help them in other nations—even U.S. cities—where they don't have a presence.

Reprinted with permission from *Principal's Report,* Institute of Management & Administration, New York.

Chapter 14

How Granite Rock Co. Won the Malcolm Baldrige National Quality Award

Bruce Woolpert

Aggregate producer Granite Rock Co. (Watsonville, California), in 1992 became the first construction-industry company to win the Malcolm Baldrige National Quality Award. In this chapter the company's president, Bruce Woolpert, tells, along with much else, how and why they empower heavy-equipment mechanics and drivers to select the firm's new equipment.

The process involves interviewing technical representatives from equipment suppliers, and doing comparison testing of the equipment. Some other companies think, "Don't ask a concrete mixer driver to order his costly equipment—you'll end up with a cab with a CD player, stereo music system, and chrome wheels." We found the opposite to be true.

Mechanics, drivers, and the crew on a paving team have a secret. As soon as they are asked to do a job that was formerly done by management, they go into a conference room, close the door, and say, "We're going to do a better job than management's ever done." Their pride drives them. I've seen salespeople come in from Peterbilt, Cummins Engine, and so forth, and be raked over the coals by these groups. They bring up every little issue with every transmission. Our experience is that their recommendation will be better than what management ever did. It may make things difficult for management, but you *must* tell them they did better, in cases where indeed they did.

How Granite Rock Won the Award

In 1992 for the first time, a construction-industry company won the coveted Malcolm Baldrige National Quality Award. Leaders of the Granite Rock Company, of Watsonville, Calif., south of San Francisco, accepted the award at ceremonies in Washington, D.C. Not primarily a contractor (one division is a general engineering contractor) but a construction aggregates, asphalt, and ready-mix concrete supplier, Granite Rock is nonetheless in the construction family. And the story behind its success holds many lessons for contractors.

Genesis of its pursuit of the top quality prize. When the two Woolpert brothers took over their family-owned Granite Rock Company in 1985, they resolved to raise the quality ante of an already strong company—it had survived nearly 90 years.

An excellent article in *Inc.* magazine (Boston) in March 1992 recited how president Bruce Woolpert began his quality quest: "He spent time in the quarry, at the [concrete and asphalt] plants, in the mixer trucks, asking people—he shuns the word 'employees'—what they liked and didn't like about their jobs. He had them list companies they themselves patronized because of their excellent service. He invited [management guru] Tom Peters in for a day-long presentation, making sure a cross-section of employees attended."

Granite Rock's marketing services manager Greg Diehl recalls that Woolpert then spelled out nine quality objectives, and challenged his coworkers to "improve quarter by quarter in everything we do."

In 1987 the quality quest took on extra urgency when several midsize competitors in Granite Rock's market area (six counties south of San Francisco) were bought out by European companies with deep pockets.

About that time, Granite Rock leaders resolved to enter the Baldrige competition. Woolpert knew his company wasn't likely to win the first year, but that didn't discourage him—quite the contrary. Explains Diehl, Granite Rock used the submission process and feedback it got from judges as a ladder of improvement. The company entered four times before it won.

Says Diehl, "The beauty is that every entrant gets back an in-depth, written critique. Basically, it's free consulting: 'Here's what we [the judges] think, and why.'

"Two years ago the judges gave us 115 suggestions for improvements. We addressed about 100 of them."

A list of all the steps Granite Rock has taken would fill many pages; the following highlights are from the executive summary of the company's entry:

- Granite Rock wanted to benchmark its concrete delivery operations against the industry's best. "When no [concrete industry] companies could be found that were also measuring on-time delivery of concrete, Granite Rock [started exchanging data] with Domino's Pizza, a worldwide leader in on-time delivery of a rapidly perishable product...[which is what] ready-mix concrete is."

    Domino's not only tracks delivery time worldwide, but local Domino's operators individually track delivery performance under traffic conditions essentially identical to those faced by Granite Rock's various concrete branches.

- "Statistical process control is in place in all three product lines (aggregate, concrete, and asphalt)." Granite Rock believes it's the first in the industry to manage using this approach.

    One example of its use: Granite Rock records variation in amount of cement in concrete batches. The coefficient of variation has been cut 50% in two years. Why bother? When Granite Rock supplies too much cement in a concrete mix, it's wasting cement (and hurting its bottom line), and when it under-supplies cement in a mix, it's cheating customers. It wanted to sharply cut both problems, and is succeeding.

(Continued)

- Granite Rock's concrete drying shrinkage is reportedly less than half its competitors'. The company attributes it to its high-quality rock and mix-design techniques.

- Granite Rock (GR) sees improving safety as a quality matter. In 1991, its workers' compensation insurance experience modifier was 0.46, meaning the company is twice as safe a place to work as the average competitor. Continuous improvement is driving more than 100 safety improvement steps each year. GR sees safety emphasis as a win–win deal: It's a way of saying to coworkers, "We put you first." And cutting accidents cuts costs, therefore improving GR's bottom line.

- Granite Rock collects receivables from 98% of its customers within 90 days; it says the regional average for its industry is 84%.

- Granite Rock predicts that facility owners will increasingly insist their contractors use performance specifications. (Most currently specify the amount of cement—rather than the concrete durability, for example.) The company conducts seminars to introduce its customers (contractors) and their clients (facility owners) to performance specifications and similar topics.

What about the future? Now that Granite Rock has won the Baldrige Award, it's not eligible to compete again for five years. Woolpert and company fear they'll be hurt by loss of the annual evaluation and feedback provided by the judges.

So they plan to continue filling out the Baldrige submission form each year, as though they were entering. They will then put together a panel of Baldrige-quality consultants to evaluate their "submission." This annual evaluation process, the Woolperts believe, will keep Granite Rock on the cutting edge of quality.

In 1994, Granite Rock won the Governor's Golden State Quality Award (California's Baldrige), and received feedback from this award submission.

How to Start a Quality Program

Granite Rock's marketing services manager Greg Diehl says any contractor can start a corporate quality program very simply. He suggests:

1. Talk to your employees.
2. Develop a list of corporate quality objectives, based on your firm's strengths, weaknesses, competitors' positions, market trends, and so on.
3. Attend the Quest for Excellence conference, featuring in-depth presentations by the latest year's Baldrige winners, which is held in February in Washington, D.C. It's being administered by the Association for Quality and Participation (Cincinnati, 513-381-1959). Hotel reservations: Washington Hilton and Towers (Washington, D.C., 202-483-3000).
4. "Read, Read, Read."
5. Enter the Baldrige. Happily, Baldrige Award criteria were updated for 1995, and are considerably easier to read and understand. Award criteria and application forms are available free from: Malcolm Baldrige National Quality Award, Building 101, Room A537, National Institute of Standards & Technology, Gaithersburg, MD 20899; 301-975-2036.

(Continued)

Diehl suggests that all contractors enter. "Don't let the lengthy and complicated entry form discourage you." Feedback from Baldrige judges can be a powerful tool in improving the quality of your services.

6. One additional suggestion: Closely study how Granite Rock won the award. The company has had so many requests for information on its quality approach, that it has installed a system that automatically faxes details to those who call a special phone number: 408-761-2300 to handle questions (it was still in service when this book went to press).

Reprinted with permission from *Contractor's Business Management Report*, Institute of Management & Administration, New York, 1993.

At Baldrige Award Ceremony, Granite Rock Employees Are the Stars

[At a February 1993 ceremony at a Washington, D.C., hotel, the latest winners of the Malcolm Baldrige National Quality Awards were honored. In the case of most winners, company leaders spoke, but in the case of one, Granite Rock (Watsonville, Calif.), the speakers were seven employees from all levels. One GR speaker, Paul Bush, provided what attendees report was the evening's highlight. He kindly gave a copy of his talk to the editor of this book. The remarks follow.]

"As you know, I am Paul Bush, and I am an employee of Granite Rock. My job is driving a concrete-mixer truck and being a driver trainer instructor. I am now enrolled in one of Granite Rock's Professional Development Programs because for 30 years I have struggled with dyslexia. This problem affects 10% to 20% of our population.

"Here's how it all started. Two years ago in a meeting with Bruce Woolpert, CEO of our company, Bruce asked me to recap, in writing, what we had talked about. I asked myself whether I should tell him I forgot my glasses—that I don't wear? Or should I say I would write it at home? Or should I tell him the truth, that I cannot read or spell? Since I didn't want Bruce to know my secret, I asked to do it at home. He said, 'I'd like to take it with me now.' I had to tell the truth. I explained I couldn't read or spell. I felt very small.

"Immediately, Bruce asked if I would like to do something about my problem. Although I didn't really want anyone else to know, I said, 'yes.'

"Three weeks later Bruce called [Bush works several miles away in Santa Clara] to explain he hadn't forgotten me, but was having trouble finding someone who teaches adults. I felt kind of sad because I had started to think about what it would be like to read and write.

"I had heard about dyslexia and that dyslexics were supposed to see things backwards and upside down. I knew that didn't apply to me because I saw the same as everyone else.

"Bruce called again to say he had found someone who's in the top of her field in teaching adult dyslexics. I was a nervous wreck. I didn't know if I was dyslexic, and I didn't want to fail another test. As it turned out, there was no pass or fail for the test that I was given. By the end of a two hour session, I learned there are three key ways that we learn, and that the way I learn is kinesthetically and auditorily, that is, through writing and motor control and through what I hear and say aloud. Knowing about my strengths made it easier to accept my difficulty remembering letters and words, which is typical of dyslexics. It is what makes it so hard for us to learn by conventional school methods. Mostly, I learned that I could learn with a different type of teaching.

"Two years later I'm still not able to write a novel, but through Granite Rock's Professional Development Program, I will be able to write one if I want to, someday.

(Continued)

> That class I attend once a week turned my life around. It has improved my confidence and convinced me that I can learn most anything. It has opened new doors to my professional and personal life. For instance, I have always wanted to be a private pilot, but felt I would not be able to get through the reading requirements of ground school. Now I know I can.
>
> "The Professional Development Program has also made me realize I have a responsibility to help other dyslexics. Less than a year ago only a few people knew I had a reading and spelling problem, and now I'm standing before you to testify to the importance of Professional Development Programs like Granite Rock's. Learning for any employee brings benefits to himself, his family and friends, his employer, and to society. I know that for me, if it hadn't been for Bruce Woolpert's persistence and Granite Rock's Professional Development Program, I wouldn't be where I am today.

The above news story lists many of the things Granite Rock did in order to win a Malcolm Baldrige National Quality Award. It does not tell why, or how they went about transitioning from a traditionally run company to one in which they partner with customers and employees. Granite Rock "empowers" those two groups. That is, it looks for ways to serve them better. (The remainder of this chapter is based on Bruce Woolpert's talk to the Associated General Contractors' 1994 convention in Orlando, Florida.)

Granite Rock before the Transformation

Back in 1986, the only data that Granite Rock (GR) collected regularly was financial. GR probably had one of the most sophisticated cost accounting and general accounting systems available, for both their materials business and their construction company, Pavex Construction Company. Nothing surprising here—this is the way most American businesses operate.

A second GR management key: They really enjoyed engineering, and they enjoyed introducing new technology to the aggregates and construction industry. Often this was done without even asking customers if they wanted the innovations. In other words, they would invent something and then try to sell it, to convince customers they wanted and needed it.

Listening to Customers

In 1986 that business approach started to change. GR people began to talk with customers—both contractor customers and homeowner customers, people building their own houses, pools: "How do you feel about Granite Rock?"

One of the things you learn when doing this is always to overstay your visit. For the first 20 minutes you will hear the things you expect: "Granite Rock supplies products that meet specifications. It does a better job than its competitors." Not unexpected—the company had been a technology leader and had been in business for over 90 years.

So then we just continue to sit there and the customer begins to open up. "Well, there is one thing we'd like to have you change." This is why you want to stay seated. At this point, the customer starts to provide the information you really need: Granite Rock is somewhat inflexible.

For example, one customer asked us to deliver concrete on a Sunday morning at 2 a.m. Instead of saying yes, we tried to talk the customer out of it, saying, "Do you really need it then?"

We had become very focused on supplying standard products with standard service. We were great only so long as you asked us to do ordinary tasks between 7 a.m. and 4 p.m. And this inflexibility concerned me: There was a danger that we would have no ongoing relationship with our customers unless we were willing to go out of our way to do what they wanted. To make things worse, we did not really believe we were inflexible.

Further, it's relatively easy for a new competitor to enter our industry. In ready-mix concrete, you can buy a plant for $80,000, and you can buy used mixer trucks for as little as $15,000 apiece. If someone asks for concrete on Sunday morning at 2 a.m., some of these new competitors will do it. Eventually you will start to lose market share, and someone who was small will grow.

The second disturbing thing we learned in 1986 was that we had become very centralized in our decision making. Bureaucratic. No one could do the simplest thing—order business cards, for example, without getting a VP's approval. The pile of phone messages on the VP's desk grew higher and higher. The executives were rushing around more and more, trying to take care of lines of people outside their offices. Bureaucracy was starting to kill us—and employees were starting to tell us this: It simply took too long to make decisions.

We did not believe that we could survive in business for another 86 years (the company was founded in 1900) with either this inflexibility or this bureaucracy. So we began reforms, starting with empowering the customer.

The challenge was to put the customer in charge. In most companies' organization charts the president is at the top, and the customer is at the bottom. It takes a long time for the customer's requests to filter to the top. So we made three changes to flip the organization chart upside down—putting the customer in the lead position:

1. *From now on, if a customer asks us to do something special, the answer is, "Yes, we'll do that." The only exception will be if a customer asks us to do something illegal or immoral.*

Today at Granite Rock that concrete gets delivered at 2 a.m. on Sunday. (The customer used it to construct a narrow levee near San

Francisco Bay.) To do it, we might have to add backup lights on our mixer trucks, and go out on the levee during the daytime and practice backing up, to be sure we made no mistakes at night.

2. *We began a series of customer surveys, although it took awhile to make them truly effective.*

Our first surveys were right out of the textbook. We asked, for example, "Rank the importance of purchase factors (and we listed several); and "Describe the performance of the three suppliers you use most often." In other words, for a mathematician (my background is in math), they were beautiful: The surveys were free of bias and ambiguity. However, we found that we needed more. We were getting what people *think,* but not what they *feel*. A thinking thing is, "Granite Rock meets specification." In other words, I can compare your results with this project spec, and can determine when you passed. The feeling stuff is, "When I had a problem, GR stood behind me. I didn't feel abandoned. I feel you treated me fairly."

I believe customers make decisions about who to contract with, and who to buy construction materials from, 60 to 70 percent based on positive feelings.

Our surveys were yielding only thinking answers, not feeling. Then, in one of our quality meetings, a mixer-truck driver said, "My experience in school was a pretty emotional [feeling] thing. I think if we switch our survey scoring system from 1 to 10, to one where we use grades A through F, we'll get at the emotional content of our customer relationships."

We tried it in one city, it was tremendously successful, and now we use it across the entire company. It turns out that when customers give you an A, the contractor feels, "I give very few of these A's." Sometimes we get an F, with powerful [expletive deleted] comments. You know there's feeling involved there! Figures 14.1 and 14.2 show the firm's questionnaire, sent to aggregate customers and asphalt customers, respectively.

The next step, of course, was to figure out why we get the A's, and do more of that. And why the F's, and stop doing that.

3. *We empowered the customer to decide whether or not we earned the money we charged them.* We do this through the Granite Rock Short Pay Method (see Fig. 14.3). The back of our invoices says to customers, "If you are not satisfied with the Granite Rock service you just received, then don't pay us for it." And customers need not justify not paying us. They can just scratch out the line items they don't like on their bills, subtract those items from the total, and pay the rest (see Fig. 14.4).

We did this because 90 percent of all customers don't complain. Everyone knows this: You're in a restaurant, the service has been lousy, the food is mediocre. You get your bill, go up to the cashier and the attendant says, "How was breakfast?" To which you reply, "Just

GENERAL COMMENTS

Here is what I like best about Graniterock: _____

_____

_____

What aspects of Graniterock's service or quality of products need improvements? _____

_____

_____

Are you aware of Graniterock's Delivered Aggregate Program? ☐ Yes ☐ No

Comment: _____

_____

_____

_____

Your Name: _____
(optional)

Company Name: _____
(optional)

Phone Number: _____

Please call me: ☐ Yes ☐ No

Regarding: _____

Thank you! Please fold with stamp showing, staple and mail.

Graniterock
P.O. Box 50001
Watsonville, CA 95077-5001

▶ **Graniterock**

Dear Customer:

We, at Graniterock, thank you for your business. We are again looking for your input to help us learn how to serve you better. Your response is important to us and will remain strictly confidential.

This survey is for aggregate products and our transportation services. These products include fill material, sand, coarse aggregate, base rock and drain rock, and trucking done by Graniterock trucks. Your replies should refer to these products and services only.

When finished, please refold this survey, staple it, and drop it in the mail. It is already stamped and addressed to Graniterock. What you tell us will be used to improve the products and services supplied to you.

Please enjoy a cup of coffee or a soft drink on us while filling out the survey. Thank you very much for taking the time to give us your feedback.

Hal Poulin   Mike Marheineke

General Manager   *General Manager*
Aggregate Division   *Transportation Division*

Figure 14.1 Customer questionnaire mailed yearly to customers of Granite Rock's quarry operations.

Please rate the **customer service** you receive at **Wilson Quarry:**
(Check the appropriate box)

	No Contact	The Best	Better Than Competition	Average	Below Average	The Worst
Weighmaster/Scalehouse						
Dispatcher/Scheduler						
Loader Operator						
Sales Representative						
Plant Manager						

Comments:

Please rate the **customer service** you receive at **Southside Sand & Gravel:** *(Check the appropriate box)*

	No Contact	The Best	Better Than Competition	Average	Below Average	The Worst
Weighmaster/Scalehouse						
Dispatcher/Scheduler						
Loader Operator						
Sales Representative						
Plant Manager						

Comments:

Please write in the names of the suppliers you use most often for **Aggregate Products.** *Then grade each company using this scale:*

A = The Best
B = Above Average
C = Same as Competition
D = Needs Improvement
F = Terrible

	#3 Supplier	#2 Supplier	#1 Supplier (used most often)
Dependability			
Product Quality			
Product Availability			
Meets Job Specifications			
Plant Availability			
Pricing			
Resolves Problems Quickly			
Ordering Convenience			
Billing Accuracy			
Fast Loadout Time			
Product Knowledge			
Eager to Help			
Responsive to Special Needs			
Full Loads/No Overloads			
Credit, Payment Terms			

What would you like to see (products or services) from aggregate suppliers that is not currently offered? ______

If you do not use Graniterock's Transportation Division for trucking, skip this portion of the survey.

If you use **Graniterock's Transportation Division,** please rate the **service** you receive compared to the **average** supplier:
(Check the appropriate box)

	No Contact	The Best	Above Average	Average	Below Average	The Worst
Weighmaster/Scalehouse						
Dispatcher/Scheduler						
Loader Operator						
Sales Representative						
Plant Manager						

Comments:

Please write in the names of the **trucking companies** you use most often: *Then grade each company using this scale:*

A = The Best
B = Above Average
C = Same as Competition
D = Needs Improvement
F = Terrible

	#3 Trucker	#2 Trucker	#1 Trucker (used most often)
Do your orders arrive on time?			
Are the dispatchers responsive to your needs?			
Is it easy to place orders or requests with them?			
Are their prices competitive?			
Are their dispatchers helpful?			
Are their drivers courteous?			
Do they address problems fairly and quickly?			

Figure 14.1 *(Continued)* Customer questionnaire mailed yearly to customers of Granite Rock's quarry operations.

Granite rock

Dear Customer:

Every year we send a mail survey to Asphaltic Concrete and Road Oils and Emulsions users throughout the Bay Area. The responses help us understand the needs and guide us in making appropriate improvements.

We need your response and comments. Please take a few minutes to complete this survey. I have enclosed a crisp new $1.00 bill for a cup of coffee or a cold drink while you take a moment to reflect on your purchases during the past year. When finished, just refold it and drop it in the mail. For your convenience, it's already stamped and addressed.

The survey is divided into three sections. Please respond to the section that best fits your knowledge and experience with your suppliers. If more than one section applies to you, answer as many as you would like.

Your response will be kept confidential. If you would like us to call you to discuss your ideas or concerns, please be sure to include your phone number in the optional section at the bottom of this page.

Thank you!

Sanjer Chakamian
Division Manager
Northern Road Materials

Which Graniterock branch do you purchase from most often? Please circle one.
South San Francisco
Redwood City
San Jose (Berryessa Road)
San Jose (Capitol Expressway)

Optional Information:
Your name: _____
Company: _____
Telephone: _____

Graniterock
365 Blomquist St.
Redwood City, CA 94063

Graniterock
365 BLOMQUIST ST.
REDWOOD CITY, CA 94063

I wish Graniterock would improve.....
_____
_____
_____
_____
_____

What aspect of Graniterock service or product do you like the most?
_____
_____
_____
_____

The least?
_____
_____
_____
_____

Please tell us about products or services you would like us to add to our facilities.
_____
_____
_____
_____

Figure 14.2 Customer questionnaire mailed yearly to customers of Granite Rock's asphaltic concrete, road oils, and emulsions operations.

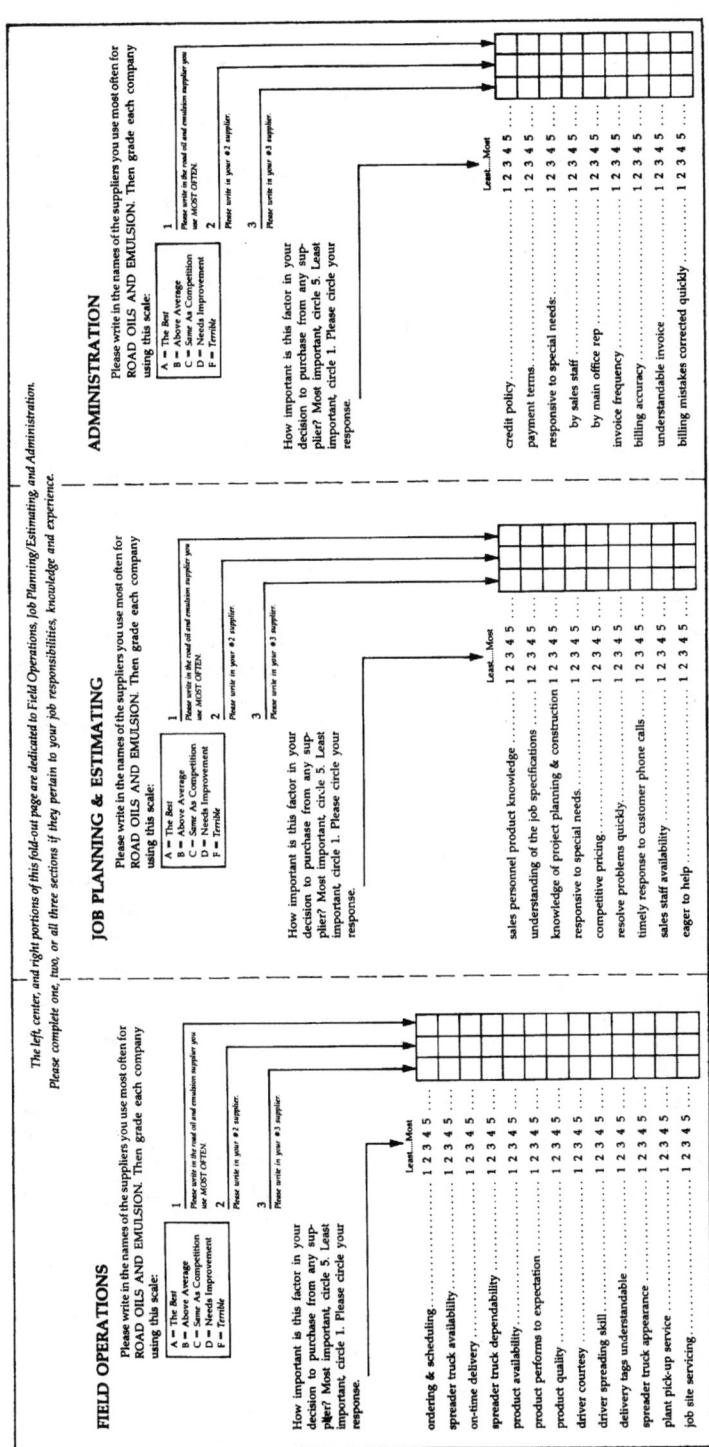

Figure 14.2 *(Continued)* Customer questionnaire mailed yearly to customers of Granite Rock's asphaltic concrete, road oils, and emulsions operations.

> **SHORT PAY POLICY**
> Our philosophy of total customer satisfaction can be summarized by the following statement:
> "If you are not satisfied for any reason...don't pay us for it."
> This means that if any part of this invoice is incorrect or if you were unhappy with the products or service received from this transaction, let us know right now. Simply scratch out the related line item, write a brief note about the problem, and return a copy of this invoice along with your check for the remaining balance. Someone will contact you immediately to resolve the problem.

Figure 14.3 Granite Rock's "Short Pay Policy," as spelled out on the back of each invoice.

fine." You walk out the door, and say to your wife or business partner, "I'll never come back here again." And so most businesses that fail don't know why. Restaurants and other businesses usually fail because people stop coming.

When we instituted the Short Pay Method, 1.6 percent of sales were held up by legitimate customer complaints. (We have not yet had any customer cheat on this.) Today, by fixing problems in areas the Short Pay Method told us needed improvement, we are down to less than 0.5 percent and falling.

Hiring Employees

While the quality-building tools discussed in this chapter can make a big difference, whom you hire is at least as important. (Transient contractors have an advantage here—many travel around the country, hire a crew just for one job, and move on. So it will be easier for them to adopt the following ideas than it would be for most companies, where employees remain for years.)

Hire people who believe as you do. Southwest Airlines came into San Jose and pushed out American Airlines—a far bigger company. Southwest hires based on attitude more than on knowledge and skills. They found that you can teach people job skills, but you cannot teach people to be sensitive to customer needs, or to have fun with customers. So hire people for good habits and attitudes, as much as or more than for their skills.

Empowering Employees

Once we were empowering customers, empowering employees had to follow, to support the "Yes, we will" commitment. When workers graduate from GR's apprenticeship program, or from one of the unions' programs, they have just completed a four-year, post-high school pro-

					I N V O I C E						
BRANCH - 331 SALINAS BRANCH (408) 424-0511										GRANITE ROCK CO. P.O. BOX 50001 WATSONVILLE, CA. 95077-5001 (408) 724-5611	
SOLD TO: SALINAS CA 93908						CUSTOMER NUMBER		PERIOD ENDING 02/18/95		MAIL PAYMENT TO:	INV. NO. NOS - PAGE 1
						P.O. JOB NO	238				
TAG NO	CAR NO	DATE	DESCRIPTION	QUANTITY	UNIT	UNIT PRICE	QTY x PRICE	DISCOUNT	SALES TAX	NET AMOUNT	GROSS
03731 60-5-500		02/10	5.00 SACK CONCRETE/60-5-500 3/4" ROUND ROCK/EXPOSED MIX	9.000	CU YD	65.000	585.00	9.00	38.03 37.44	613.44	623.03
03741 60-5-500		02/10	5.00 SACK CONCRETE/60-5-500 3/4" ROUND ROCK/EXPOSED MIX	.500	CU YD	65.000	32.50	.50	2.11 2.08	34.08	34.61
			PRODUCT TOTALS	9.500 *			617.50	9.50 *	40.14	647.52 *	657.64 *
			INVOICE TOTALS SALES TAX % 06.50					9.50 *	40.14 39.52	647.52 * —34.08	657.64 *

Late delivery on last truck. → 613.44

Payment Enclosed → 647.52

CUST	INV # NOS - PERIOD ENDING 02/18/95	If payment received by 03/10/95 pay	657.64
SALINAS CA 93908 BRANCH - 331 SALES TAX % 06.50 TOTAL DISCOUNT INCLUDED IN NET $10.12		If payment received after 03/10/95 pay A service charge will accrue on past due balance	
	RETURN THIS PORTION WITH PAYMENT		

Figure 14.4 A Granite Rock invoice, in which the customer has taken advantage of the company's short pay policy, and not paid for one item.

gram. They are told they are journeymen or some such. That is, we tell them they know everything they need to know.

That's a lie. The mechanics in our shop were not up to date in how to repair electronic transmissions and electronic equipment on our Cat loaders and other equipment. They all wondered if they themselves were defective, because they knew they didn't know things. And people were feeling professionally old in their 40s. We needed lifelong learning as a company-wide value.

And we wanted coworkers to feel that they "own" their jobs. Sometimes there is confusion about who is responsible for improving a job. If I am a payroll person in Granite Rock, do I wait for my supervisor to come up with ideas on how to improve payroll processing, or do I do it myself?

We told all our coworkers that they "own" their jobs, and further, that doing a good job requires two things: doing their job, and improving it.

The reaction was wonderful: "Thank God, I get to do something about these problems that no one was paying attention to."

We developed baseline goals, and each year we require improvement: in customer satisfaction, in our safety program, in product quality, and in other aspects.

As a way of saying to our people, "You come first," our corporate objective for safety was written thus: "Safety shall come before all else, shall come before meeting production goals, shall come before meeting customer commitments, shall come before doing anything, shall come first."

Employees read that and said, "Oh, that's nice." Then we issued a baseline goal of zero accidents, and said that managers would be measured against that baseline, and the first reaction was, "I don't think we can achieve that."

Then our people started to think, "What would happen if we put safety ahead of all else? Could we get to zero? Oh, I start to understand the objective now. You really mean, put safety ahead of all else? With that, maybe I could get to zero accidents."

I really did not think, in GR's early years of implementing total quality approaches, that we would make as much progress as we have. But as people have become empowered, they feel complete job ownership, and they have had so much training, they now know just as much as management does.

Shifting Middle Managers from Being Bosses to Being Coaches

Of about 60 managers, we had about three who had great difficulty making the transition from manager to a more coaching type of leadership—that is, letting others decide, make decisions. We saved about two and one-half of the three troubled managers in this way:

We sat down with them and said, "You became a manager because you were very strong at your work. You became maintenance supervisor because you were a great mechanic. How did you become a great mechanic?"

And the answer always was, "I worked for this person who let me make decisions, taught me a lot."

To which we'd reply, "Is that the kind of manager you are now? Are you a controlling kind of manager, or the kind you liked in your boss?" You have to go face to face with them.

Some people, when they become supervisors, become directive. They think, "I'm the king, I'm in control, so I'd better show people that." Much of the reason for this is that they think the president of the company wants them to look like they are in control. How many times have you been in a company where the president calls on the phone and asks a manager a question. And the manager feels he should know the answer. So he thinks, "For that to happen, I'd better keep tight control of things." To make sure that doesn't happen, I think we presidents, when we make such calls, should say, "This is what I'm interested in; can you tell me who to talk to?"

Employee Training

At most companies, most knowledge is hoarded by top management. When I was at Hewlett Packard, as a member of management I was one of those who went to the conventions and seminars, who got the magazine subscriptions. Rarely would others have the chance to be enriched in these ways. At Granite Rock this top-only philosophy has been changed.

And we have made a huge investment in training—50 in-house classes in asphalt mix design, concrete mix design, concrete finishing, and many other areas.

We also found that our people's understanding of the science of concrete was sketchy. I would ask, "What is water–cement ratio?" "Why is it important?" And I would get the wrong answer more than I would like.

So for each coworker we developed an Individual Professional Development Plan (IPDP). We got rid of our performance evaluation system—because it looks backward—and decided to look forward only, by helping each coworker develop his or her talents. Individuals are allowed to decide where they want to go, where they believe their skill deficiencies are. They write down what they believe are their skill deficiencies, and they review this list with their manager with the idea of enhancing the plan.

After IPDP had been in place for three or four years, the trust level had risen so much that people really began to open up. Someone wrote, "I'd like to learn how to read better." We asked the manager, "Could you find out what he means by reading better?" This person meant, "I can't read at all."

We found out that 16 percent of men and 8 percent of women have difficulty reading and doing math. Yet many people with reading disabilities are highly intelligent. Einstein had dyslexia. We have our geniuses, too.

The Granite Rock employee who couldn't read was a superb mechanic—he could fix anything in the quarry, but in his 16 years with us he had not been able to read one word (see box on page 214–215).

We now have four men and one woman (of 400 employees) in our reading program, every one of them with more than 10 years' experience at Granite Rock. In every case, no one knew they had a reading disability. And all five of them are emerging as more powerful and dedicated people.

The average Granite Rock person receives 37 hours a year of training, and the company spends $2000 a year per employee on training.

The result of empowering clients and empowering our people? GR's market share is up 88 percent, to nearly double its 1986 level.

Measuring Performance, and Setting "Stretch Goals"

In the beginning, in 1987, we measured 15 aspects of company performance and set baseline goals for improvement. Now we have 66 measures.

We asked contractor-customers who bought our concrete, "What's the most important thing to you in the concrete you buy?" The usual answer was "On-time delivery." And for 86 years we had believed that the most important thing to them was the quality of our concrete.

So we decided to measure our on-time delivery. We guessed that we were delivering on time—that is, when promised, plus or minus 30 minutes—about 90 to 95 percent of the time. The truth turned out to be 68 percent, about the same as our competitors.

Today, by using computers and exchanging ideas within Granite Rock, and by empowering dispatchers and drivers, we are at 94 percent on-time delivery.

Because our customer surveys say they want plus or minus 15 minutes, not 30, that's the new measure we have adopted.

One tool we used for improving was to benchmark against the performance of the best. We chose Domino's Pizza, because they have a 50 percent market share in their business—basically because of the 30-minute promise. Quality of pizza does not matter as much as on-time delivery. By studying how Domino's does it, and why, we learned some things.

The same thing happened with our Nova award-winning GraniteXpress aggregate-dispatching system. We asked customers for our crushed rock and asphalt, "What's the number one thing you want?" The answer was, "I want our trucks to be in and out of your yard fast."

So we adapted to rock dispensing the idea of the automated money-dispensing ATM machines pioneered by banks. A rock customer's truck driver pulls up under our rock bin, inserts an ATM card, keys in the product wanted, gets it, and drives out. The system works 7 days a week, 24 hours a day.

In California quarries the average loading time is 24 minutes. GraniteXpress is down to 7 minutes. The California Public Utilities Commission's trucking rate is $62 a hour, so at $1 a minute, $17 is saved on each aggregate load.

We are also more accurate in giving customers what they ordered: When we were loading the old way, we averaged 800 lb underweight. Today we average underweight only 60 lb.

Another example is rework on newly placed paving. One of our businesses is as a general engineering contractor. For paving, we would budget a certain amount of money for rework—grinding to meet the smoothness spec. Then we started training our construction crews in the right way to do things, and started measuring miles of roadway paved without grinding. We have cut paving rework by 90 percent.

Finally, GR's labor productivity is 30 percent higher than industry average. How do we do it? The same way every company that excels at quality does it—by eliminating errors. Quality guru Philip Crosby and others have written that the equivalent of 25 percent of the American workforce is needed to correct others' errors. What we have done is find ways not to make those errors in the first place, rather than correcting errors. And I challenge contractors to do the same—look in your companies, and find nonproductive activities.

Why 50 Percent of Total Quality Management (TQM) Programs Fail

One of the great things that happens when you are a Baldrige award winner is that you get to talk with other Baldrige winners. You see commonalities. I have learned five things to do—and their opposites to avoid:

1. *Focus on the customer.* Focus your financial investment and your time investment on the customer. The customer *must* be in charge. If

you don't do this, particularly in large companies, internal management becomes more important—most employees see their job as pleasing management, not customers.

At such a firm, ask someone, "What's the most important thing to do to advance your career?" Answer: "Make my boss happy."

If your focus is on customers, you will find out what makes them happy, and do more of that. And your market share will rise. Ours has.

2. *Avoid TQM language lessons.* Avoid emphasizing TQM terminology—terms such as root-cause analysis and statistical process control—too early in the cultural change. We use these terms, but only as tools—we don't treat them as our gods.

Some people who adopt TQM do. Wrong. If you do, your workers will think they need to learn a new vocabulary—that there's something defective about the way they think. They shrink. They feel less powerful. So start out by saying, "From now on we're going to focus on customers." Once coworkers have assimilated that, new TQM tools and terminology can be introduced gradually.

3. *Avoid the Messiah effect.* The Messiah effect comes about like this. Granite Rock hires a new VP for quality, and the president calls everyone to a meeting. "Today I'm pleased to announce I've hired a new vice president for quality. He or she comes from IBM (or some other quality-leading company), and we're lucky to have this person here to lead us."

So everybody waits a year to see what the new VP–Quality does. A year later nothing has happened, so the president has a career-counseling session with the VP–Quality about how he or she is not working out, chemistry isn't quite right, and the VP leaves. The lesson is that quality leadership cannot be delegated—it must come from the CEO.

4. *Trust customers and trust employees.* Some people believe that you cannot trust customers. I have learned that this is true only if the relationship has already been destroyed—then the customer may try to "cheat" you to get even.

Second, trust your employees. We trust ours at Granite Rock—for example, quality teams of drivers and mechanics together decide which heavy equipment GR will buy. The process involves interviewing technical representatives from equipment suppliers, and doing comparison testing of the equipment. Some other companies think, "Don't ask a concrete mixer driver to order his costly equipment—you'll end up with a cab with a CD player, stereo music system, and chrome wheels." We found the opposite to be true.

Mechanics, drivers, and the crew on a paving team have a secret. As soon as they are asked to do a job that was formerly done by management, they go into a conference room, close the door, and say,

"We're going to do a better job than management's ever done." Their pride drives them. I've seen salespeople come in from Peterbilt, Cummins Engine, and so forth, and be raked over the coals by these groups. They bring up every little issue with every transmission. Our experience is that their recommendation will be better than what management ever did. It may make things difficult for management, but you *must* tell them they did better, in cases where indeed they did.

5. *Don't think of TQM as an add-on.* Add-on programs are death to companies. Total quality approaches are like oil, and the old American way of running business is like water. They don't go together. Put them together and someone will get hurt, people will get angry, and things will blow up.

If you elect to pursue total quality, change your entire approach. I don't mean change everything at once—it took us seven years. Perhaps you can change only one thing—say, safety—first. But philosophically, you will have changed 180 degrees. Consider the following example of a failure resulting from treating your quality program as an "add-on":

You are faced with a promotion dilemma. One manager is recommending John for the position—he has had no complaints, his customers love him. We hear no bad things about him.

Another manager advocates that Susan get the promotion. She has really gone after total quality programs: She surveys customers, even internally. She encourages people to complain. She has received many complaints, and changed many things, and in the process has ruffled some feathers. Simple little things, like changing a form that someone in accounting developed 15 years ago, and "that was my form and it was good 15 years ago, so it should be good today, right?"

Depending on your scoring system, both are good people. Whichever you promote, you lose the other one. So switch cold turkey to the new approach, or don't bother.

Chapter 15

Why One Contractor Calls Partnering "Project Quality Planning"

Ronald L. Deffenbaugh

As partnering is to the project, total quality management (TQM) is to the construction company. In both cases, operations are improved through teaming, joint setting of goals, and further systematizing of operations. Ron Deffenbaugh, quality VP at McDevitt Street Bovis (Charlotte, North Carolina), one of the most successful contractors in the use of TQM, gives two examples of how they apply it.

1. On the new office building for the Comptroller of the Currency (Washington, D.C.), mechanical contractor John J. Kirlin and electrical contractor Dynalectric collectively designed a process in which mockups of mechanical and electrical systems were used to reach consensus on acceptance prior to mass installation. Before construction began, the team identified and corrected excessive duct leakage and possible system interferences. Rework and punchlist headaches were reduced.

2. Before construction begins, we assess the adequacy of work skills of subcontractors. We do not believe that people show up to work to do a bad job. But sometimes they do not know how to do it right. In order for floors to be flat, submittals to be accurate, and government requirements to be met, people must be trained. This training is not delivered only by us. We involve the owner, design team, and especially subcontractors in training each other. By doing so, they understand the important aspects

of one another's work. We also believe that the worker who grows to do things right through training will be more satisfied.

In 1991 the Associated General Contractors (AGC) of America published their now-famous booklet entitled *Partnering—A Concept for Success*. It defined construction partnering as a "process...to establish working relationships among the parties (stakeholders) through a mutually-developed, formal strategy of commitment and communication." Other groups, such as the Construction Industry Institute, the Associated Builders and Contractors, and even the American Institute of Architects, followed with their own definitions, but our concern was with this first one that we noticed.

Our concern was this. Since mid-1988 our company, McDevitt Street Bovis (MSB) of Charlotte, North Carolina, had been committed to a process of total quality management. Our mentor was Florida Power and Light (Jacksonville, Florida), the only company outside Japan to win that country's quality award, the Deming Prize. Their subsidiary, Qualtec, a quality consultant and also of Jacksonville, taught us their quality improvement process and taught us to teach it. Part of their process was quality planning, which we had begun to utilize on our construction projects. Thus, it became known as "project quality planning." Thus, when the AGC publication arrived, the question arose, "Do we change the name of our process to partnering or continue to call it project quality planning?"

As we looked at our process, we saw it conforming to many of the key elements of partnering as defined by the AGC:

AGC Element	MSB Process
Equity	We set a ground rule for our teams that there is "no rank in the room."
Trust	We used consensus exercises to show the benefits of trust at the outset and reinforced decision making by consensus throughout the project.
Development of mutual goals/objectives	We arrived at a common mission for the project.
Implementation	We identified key tasks to promote progress toward the mission.
Continuous evaluation	We established a project lead team to monitor progress throughout the project.

Even with these similarities, however, we chose to be different and remain different by retaining the name of our process, "project quality planning." Why? Because we firmly believe that any so-called partnering effort needs to be a part of a larger, all-encompassing TQM process. In a sense, partnering is to the project as TQM is to the construction company. It simply needs to be the way we do business on every project. In this chapter, we will support this belief by taking a detailed look at the three words which comprise the name of our process.

Why Project?

Our discussions on implementing TQM on construction projects began while our company was implementing quality planning as prescribed by Qualtec. It included establishing a mission and a vision for the company and setting forth fundamental and detailed objectives to support progress toward the vision.

At a 1990 dinner in Rockville, Maryland, we began to compare a construction project to our company. The similarities were obvious. Both involved a diverse group of people joining together in pursuing a common objective. Although many companies are represented on a construction project, the objectives were seen as generally the same as those of our company—do things safely, on time, within budget, and in a quality manner, so that everyone can win together. We arrived at the obvious conclusion: If quality planning works in a company, the same approach should work on projects.

To dig deeper into the origin of the process called quality planning, one must study the writings of the quality guru, Dr. Joseph M. Juran. In his book, *Juran on Planning for Quality* (Free Press, Division of Macmillan, New York, 1987), Dr. Juran states, "The purpose of quality planning is to provide the operating forces with the means of producing products that can meet customers' needs, products such as invoices, polyethylene film, sales contracts, services, calls, and new designs for goods." Isn't it interesting that all of the items he mentions are parts of a typical construction project? So projects are like companies in that they are all different, they have both internal and customer/supplier relationships, and they can begin an intense focus on quality at any point in time. Consider these examples.

Our very first project quality planning session took place in the construction of the new office building for the Comptroller of the Currency in Washington, D.C. On this $27 million project, the session was held not at the beginning of the project but shortly after topping out of the building structure. The team, which included mechanical and electrical subcontractors John J. Kirlin and Dynalectric, formulated a mission statement which said, "To provide an optimized installation of the designed mechanical and electrical systems by realizing zero rework, minimum punchlist, and efficient start-up and turnover."

The team went on to collectively design a process in which mechanical and electrical system mockups were used to reach consensus on acceptance prior to mass installation. The team identified and corrected excessive duct leakage and identified possible system interferences up front, thus reducing rework and the punchlist.

Looking back on the process, Barney Silver, senior construction manager for the building developer, Boston Properties, said, "The results have not only been the smooth turnover of systems but more importantly the establishment of a sense of teamwork of all parties critical to the construction process." The mechanical subcontractor was pleased as well. Dan Liscinsky, vice president of John J. Kirlin, said, "It shows that preplanning is the key to quality—get it right the first time."

Mr. Liscinsky's quotation points to a key to success of project quality planning. It can be even more effective when done before construction begins. Our team at Clarksville Memorial Hospital in Clarksville, Tennessee, found this to be the case. Their mission was "to provide Clarksville Memorial Hospital with quality health care design and preconstruction services, meeting needs and expectations of all team members, using the principles of teamwork and partnering to ensure a timely construction start within budget."

By pursuing quality at this stage of the project, the team worked together through the contract document stages and met the original budget in spite of numerous scope enhancements. By involving all team members in the review of mechanical proposals, the owner not only selected the best, bondable mechanical subcontractor, but saved $200,000 on this $5 million project.

Unlike the partnering process AGC described, which starts at the prime-contractor bidding phase, we found that project quality planning is beneficial starting at any point in the project. A project team may decide to come together at any of the four design stages (conceptual, schematic, design development, or construction documents), or at any time during construction (ground breaking, precontract, postcontract, topping out, or close-out). Once started, however, the process must involve everyone, and it must be continued to the project's end so it can be evaluated at that time for lessons learned.

Since every contractor is identified by its collection of projects, it is logical that the word "project" be included in the name of the process.

Why Quality?

In the modern-day quality movement in the United States, one name stands out—Dr. W. Edwards Deming. Although supporters of the term "partnering" may connect its origination to Deming, his writings disagree with the term being used exclusively.

Multiproject strategic partnering refers to point 4 of Deming's famous Fourteen Points: "End the practice of awarding business on price tag alone." Dr. Deming's idea was to "seek the best quality and work to achieve it with a single supplier for any one item in a long-term relationship." That is an idea that is probably embraced by all of us on the supply side, but it will not happen by itself. Deming's view of quality was holistic, and he believed that adoption of all of his fourteen points is necessary to anticipate the needs of the customer, enlist the cooperation of employees, and constantly improve processes and products. Our view of TQM in construction is the same.

Since our quality mentors, Qualtec, base their process on Deming and Juran, what we learned from them was holistic as well. We learned a quality definition that said: "Quality is satisfying the needs and reasonable expectations of the customer." We committed to living the four quality principles we learned:

1. *Customer satisfaction*—Having the attitude that puts the needs of the customer first.
2. *Plan–do–check–act*—Focusing on continuous improvement by planning what to do, doing it, checking what was done, and acting to improve.
3. *Management by fact*—Collecting objective data and making decisions based on that information.
4. *Respect for people*—Supporting the capacity everyone has for self-motivation and creative thought.

We learned how to determine valid requirements in creating a win/win agreement between customers and suppliers by using the acronym RUMBA:

- *Reasonable*—They are within acceptable guidelines.
- *Understandable*—They can be understood.
- *Measurable*—Objective measurements can be taken.
- *Believable*—They are agreeable to others.
- *Achievable*—They can be met.

Beginning with that foundation, we arrived at our company mission: "To provide quality construction services for select clients." When it came to implementing TQM at our construction projects, we decided to walk our talk and include the word "quality." This meant that we would need to align our actions with our principles. We have done that in the following ways.

Customer satisfaction

We developed two ways to obtain feedback from our customers. First, we implemented a bi-yearly formal client survey, administered by an independent consultant so that it was totally confidential. That approach has allowed us to identify big-picture items and monitor our long-term improvement. Second, we implemented a bi-monthly performance evaluation with our clients in order to respond to their needs more quickly. In this process, the division manager or a designee interviews the client. Specific ratings are given on factors such as attitudes, responsiveness, and listening. Results are trended over the life of the project. We also have conducted formal subcontractor surveys and implemented on-site suggestion systems to meet our suppliers' needs. Our focus on customer satisfaction results in this type of feedback:

> We have found that McDevitt Street Bovis is committed to meeting its customer's expectations and embracing quality management, whether it's a negotiated deal or competitive bid. That impresses us. After you've won the job, you sit down with the client and subcontractors at the same time to ask "What are your expectations? Let's set some goals together for this job." We've seen you provide that leadership, and there have been fewer problems all around. It's been a win–win solution. As a result, McDevitt Street Bovis has won more work from us and we like being your client.
>
> *Bill Lee, Chairman and President*
> *Duke Power Company—End User*

Plan–do–check–act (P–D–C–A)

We created quality improvement teams on the project site to improve processes. One of the early successes was The MOB Team at the Centennial Medical Center in Nashville, Tennessee. Their identified focus was the medical office building close-out process for this $15 million project. Their mission was "to develop a process that will meet the reasonable needs and expectations of Centennial Medical Center, the physicians and their staffs for the suite occupancy process." The team, which included the owner's representative, worked through the following priorities:

Priority 1	Development of close-out documentation—included development of operation and maintenance information for all mechanical and electrical equipment
Priority 2	Training of owner personnel—developed, scheduled, and carried out training for the owner's maintenance staff
Priority 3	Punchlists, inspections, and verifications—developed, scheduled, and carried out joint inspections to facilitate final acceptance
Priority 4	Transition to tenant occupancy—created process to preplan move-in at 120 days prior and 60 days prior

By creating this formal process and working through the plan, the team was able to move 23 doctors' groups into 200,000 ft^2 of space in a period of 2 weeks with zero complaints.

Wayne J. Buck, Vice President of Support Services, Centennial Medical Center, Nashville, Tennessee, had this to say: "This project was no small challenge for us, and it was refreshing to work with a construction team that truly strived to understand and meet our needs in the process."

The MOB process is now being replicated successfully on other projects in the company.

More recently, a team called the Magicians of Coordination at the $87 million Healthcare Finance Administration project in Baltimore, Maryland, also showed the benefits of a team focus on P–D–C–A. The team includes mechanical and electrical design engineers from RTKL and subcontractors, John J. Kirlin and Enterprise Electric, as well as MSB personnel. As the team meets, mechanical and electrical shop drawings are "preapproved" prior to completion by discussing the design, thus reducing the cycle time for the submittal process. Moreover, by meeting together as the mechanical and electrical systems are designed, the team has avoided over 150 possible field conflicts. The savings for "doing it right the first time" has been estimated at $300,000!

Management by fact

In the past, measuring on construction projects consisted primarily of two parameters: cost and schedule. With a TQM approach, however, we found that, to monitor progress and thus maximize chances of a successful project, much more needs to be measured and displayed. For instance, by measuring the following processes, we develop leading indicators to monitor progress on the overall schedule:

- *Request for information process*—number submitted with solutions, number submitted versus number returned on time
- *Submittal process*—percentage submitted on time, and returned on time
- *Payment process*—days overdue by owner, days overdue by general contractor
- *Change order process*—percentage of time the pricing turnaround was on schedule, and accuracy of pricing

Measurements such as these support the P–D–C–A process by allowing project team members to be proactive in addressing problems. In fact, measuring of the submittal process has led to a new approach to procuring precast wall panels for projects. Traditionally, shop drawings are purchased with the precast material. Consequently, details are not designed until the material is purchased. Our measurements told us that there is a better way. With the agreement of the owner, we have purchased the shop drawings separately, so that the details can be designed along with the design drawings. This can typically save 2 to 3 months on the project.

Measuring also enhances teamwork, since it eliminates surprises. With all the data openly communicated or displayed, the team takes ownership and strives to improve together. We know that no one likes to be policed, so we recommend that those closest to the process do their own measuring. We as general contractor periodically verify the measurements.

Respect for people

Respect for people means much more than being nice. It means appreciating diversity and realizing that it is our diverse experience and creativity that leads to more synergistic results. Brains have been issued fairly—one per person—so we try to use every one.

We begin by involving diverse groups in our initial planning sessions so that expectations and ideas are widely communicated. In the evaluations of these meetings, we find that the participants appreciate this type of exchange because it builds a foundation of trust.

But it cannot stop there. We continue to clarify those expectations throughout the project. On larger projects, we have also extended this type of communication to on-site workers via an on-site suggestion system. We want to be sure that barriers are removed that prevent workers from doing the best job they can. Whether they ask for better lighting or better plans, we want to respect their wishes because they are the ones putting the work into place.

When work is done right, we positively reinforce it with recognition. As reported in the *Harvard Business Review,* second only to achievement, recognition is the greatest factor leading to employee satisfaction. So we try to catch people doing things right.

We typically recognize a "quality crew of the month" on our projects by rating crews against a standard of productivity, safety, cleanliness, and quality of work in place. On some projects, individual suggestion and quality-alert submitters are recognized as well. Problem-solving teams are recognized at our annual Quality Day celebrations.

These are just some of the methods we use to show our respect for people. It is most important to remember to "Just do it." For most of us in the construction industry, our only asset is people. We learned early in our quality journey how important this is when a subcontractor said this:

> I like the quality awards given on One Independence Square. And the quality team actually improves productivity on the project. I can tell by my labor reports.
>
> J. Brian Burns, Vice President/Branch Manager
> Dynalectric—Subcontractor

In summary, in order to be true to our mission of providing quality construction services for select clients, we must strive to live the four quality principles on every project. Partnering is not a selective methodology used on special projects, but a part of the way we do business.

We began this paradigm shift on pure faith, but research has supported our direction. In their 1991 report, "In Search of Partnering Excellence," the Construction Industry Institute stated that "organizations considering partnering relationships should...consider total quality management issues up front." With this corroboration of our belief, we know that we are right in our emphasis on quality as a whole rather than on partnering only.

Why Planning?

Although we began our TQM journey based on a foundation of Deming and Juran philosophy, there is one other guru who has impacted our approach. That man is Dr. Steven R. Covey, author of the book, *The 7 Habits of Highly Effective People* (S&S Trade, Division of Simon & Schuster, New York, 1990), and it is his first three habits that point to our focus on planning.

In Habit #1, Covey prescribes being proactive. Although Covey applies this habit primarily to the individual, we think it also applies to organizations such as project teams. To paraphrase Covey, project teams that do not plan say things like:

- There is nothing we can do.
- They will not allow that.
- We have to do that.
- We cannot....
- We must....

Proactive teams, however, sound much different. They say:

- Let us look at alternatives.
- We can select a different approach.
- We can deliver an appropriate response.
- We choose....
- We will....

To paraphrase Covey, proactive teams accept responsibility for their own actions and don't blame others. They become exactly what they want to be—successful.

In Habit #2, Covey advises beginning with the end in mind. So many times construction projects begin with everyone simply showing up and going to work. As problems crop up, they begin to react and become defensive. There is a better way, and that is using what Covey calls "proactive muscles" to imagine what the characteristics of a successful project would be. If this is done prior to beginning work, it becomes like a rope and pulls the team toward that end. As Covey says, the product of beginning with the end in mind is a mission statement. Here are some of ours:

- Through teamwork, we will provide a facility that meets the identified needs of Holston Valley Healthcare, Inc., in a manner that exceeds safety, schedule, and budgetary requirements.
- To produce a successful project for everyone involved by continuous improvement through the constructive interchange of experience, creative ideas, and thought.
- To successfully construct Chesapeake Power and Light's (CP&L's) O&M facility in strict accordance with the plans and specifications on schedule with maximum quality, while maintaining a safe professional environment, and through teamwork to create lasting relationships with all parties involved.

Missions like these become a source of what Covey calls "power, wisdom, guidance, and security" for the team. As he adds, it is as though they have created their own constitution.

Habit #3, however, is the one that best exemplifies a continuous focus on planning. It says to put first things first and recommends using the Quadrant II Time Management matrix, which focuses one's time on things that are "not urgent but important," as Covey recommends. Here are four of Covey's recommendations under that heading which we practice:

- *Preparation*—Before key processes which impact the mission are begun, we prepare for them. We clearly define them so that everyone understands who is involved and what needs to be done. Processes that are often the subject of contention are designed by consensus before they become urgent. Some examples include the submittal process (see Fig. 15.1), the payment process, the change

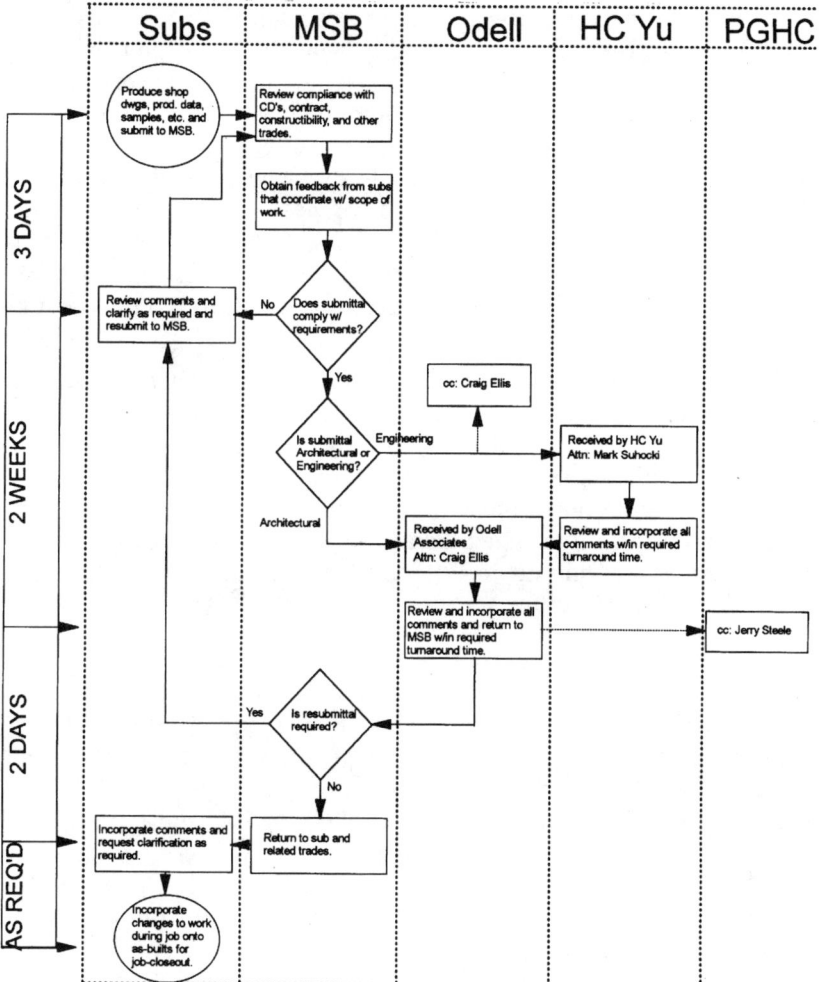

Figure 15.1 Shop drawing submittal flowchart. McDevitt Street Bovis and its partners on one project charted the process, thus providing the basis for (1) setting time limits at each stage, and (2) benchmarking if the team decides to accelerate the process.

order process, and the project close-out process. With proper preparation, future conflicts are minimized.

- *Planning*—Once our teams do their initial planning, they meet regularly as quality lead teams to accomplish the mission. This means ongoing planning. For instance, our team at the Healthcare Finance Administration Headquarters project, mentioned above, has disciplined itself to look 6 months ahead on a weekly basis. So, when the foundations were going in, they were talking about precast concrete delivery. When the topping out was taking place, they were talking about beginning tenant work. This long-term view prevents crises down the road.

- *Skill improvement*—Once processes are identified and designed, training must take place. Training is typically not urgent but very important. We do not believe that people show up at work to do a bad job. But sometimes they do not know how to do it right. In order for floors to be flat, submittals to be accurate, and government requirements to be met, people must be trained. This training is not just delivered by us. We involve the owner, design team, and especially subcontractors in training each other. By doing so, they understand the important aspects of one another's work. We also believe that the worker who grows to do things right through training will be more satisfied.

- *Relationship building*—People must know each other in order to trust one another. We make this possible at the outset through icebreaker and consensus exercises and by sharing meals together as a team. We also encourage parties to celebrate meeting goals or milestones, and sports activities where team members can form personal relationships. The payback for this is difficult to measure, but we believe it has an impact in reducing the storms that take place in the life of a team.

These three habits which focus on planning provide the direction for synergy and continuous improvement stressed in Covey's other four habits. Moreover, since planning is the first step in the P–D–C–A cycle, we believe that it is important to give it emphasis in the name "project quality planning."

Early in our quality journey, we searched for information on TQM in the construction business. That led us to a Japanese construction company that, like our mentor, Florida Power and Light, had won the Deming Prize. The company was Takenaka Kamuten Company, Ltd., and we took to heart the words of its chairman, Renichi Takenaka: "We emphasized the importance of policy management (quality planning) and upstream management—management of the starting phase of work." Like Takenaka, we found that planning is the key to achieving and maintaining quality in all phases of construction. It is therefore an important part of the name of our process.

The Results of Project Quality Planning

In other chapters in this book, the reader will see positive data from single-project partnering. While that information can be quite impressive, we know that similar successes happened prior to "partnering" when the right team met on the right project at the right point in time. Total quality management results must therefore be measured differently. MSB has had to establish high-level measurements across the entire business to determine TQM and project quality planning success.

We call those measurements "key result areas." Since project quality planning is being implemented on virtually all of our projects, its effects should be seen. Now that it has been used for about 4 years, we are beginning to see the positive results we expected.

- *Repeat business*—Approximately 70 to 80 percent of our work is with repeat clients who have been satisfied with previous results.
- *Sales*—Revenues were at an all-time high in 1993, and we believe that record will be broken in 1994.
- *Customer satisfaction*—In the division where project quality planning was first implemented, all satisfaction factors improved. The biggest improvement was in communication.
- *Employee satisfaction*—Our annual employee surveys show that the employees in the early-adopting division are among the most satisfied in the company.
- *Litigation*—From 1990 to 1994, our company's annual legal fees have dropped from $2.6 million to $600,000. There has been no litigation with the client on any project quality planning project.
- *Profitability*—Although our company changed hands in 1990 and faced a shrinking construction market and recession, we remained profitable through it all.

While we realize that a TQM commitment is a journey, we are proud of these milestone measurements and the progress they show. We do not believe that this would have been possible by partnering on selected jobs. Besides, that would not show integrity or be in alignment with our quality principles. In a TQM company, we must take the initiative and plan for everyone to have a quality success on every project. The best name for that is "project quality planning."

For MSB's outstanding quality work in 1994, it was given the U.S. Senate Productivity Awards and Maryland [State] Excellence Awards (see news release on next page).

The Maryland Center for Quality and Productivity U.S. Senate Productivity Awards and State of Maryland Excellence Awards

On October 4, 1994, Senator Paul Sarbanes and Senator Barbara Mikulski presented these awards at the Maryland Quality Conference to three outstanding Maryland organizations: the Powertrain Division of Mack Trucks, the contractor McDevitt Street Bovis, and the Naval Surface Warfare Center at Indian Head.

The award is given to private and public-sector organizations for their efforts to improve productivity and quality of their operations. The applicants must demonstrate improvements in productivity and quality leadership, human resource excellence, productivity and quality results, customer orientation, and impact on the state and local community....

The award in the service category was given to McDevitt Street Bovis, a construction services company [with regional offices] in Bethesda, Md. The firm formally instituted total quality management throughout the company in 1988, and has customized and enhanced the principles into a proprietary approach known as Project Quality Planning.

Some of the resulting improvements are a reduction in staff turnover from 12% in 1991 to 5% in 1993, a 59% [reduction] in cost of poor quality, and a 75% reduction in legal expenses as a percent of job profit. This last is an indicator of higher quality performance and better relationships with clients and other stakeholders.

There has been 100% on-time performance in project scheduling since 1989. There have been significant improvements in client satisfaction ratings versus those for major competitors in eight measured areas (initiative, overall relationship, project closeout, knowledge of the client's business, innovation, and so on). The client retention rate has increased, and in 1994, 40% of the work came from existing clients. There has been a 15% annual improvement in sales per employee from 1992 to 1993.

In receiving the award, Jeff Arfsten, Senior Vice President and Division Manager, said "McDevitt Street Bovis is honored that our unique quality and productivity practices have been singled out to receive this important award. Our belief in and practice of quality principles has been a tremendous asset" to the firm. (For information on these awards, call 301-405-7099.)

Chapter

16

How Partnering and Empowerment Made an MBE Effort a Winner

Lorry Bannes

Lorry Bannes, civil engineer, contractor, and now president of the Bannes Consulting Group (St. Louis) describes the creation of the Courtney Health Center, an inner-city health clinic the construction of which 22 years ago was done primarily by minority contractors. It was built in budget and on schedule; more important, a number of minority subcontractors grew enough on this job to be "mainlined," successful on their own. How?

Bannes set up a minority contractor development and assistance office near the site, where he taught quantity takeoff, project scheduling, component phasing, and trade scheduling. Plan and spec technical requirements and design intent were clarified.

The principal benefits from these efforts came in the areas of general requirements and conditions. Bannes had seen that young minority contractors usually performed very well when conditions permitted and even encouraged good work. On this job he tried to provide such an environment.

During the analysis and bidding periods, he talked about "how to build," including equipment sharing, tool and supply requirements, manpower leveling, and resource phasing and scheduling. He encouraged the contractors to study the history of their fee income, and to track their main-office costs and capital requirements.

How Do Your Leadership Standards Compare?

You grow leaders partly by teaching, partly by trying to be an effective leader yourself, according to Lorry Bannes, a St. Louis civil engineer, a former contractor, arbitrator, and A/E/C management consultant. Speaking before a Project Managers' Association (APM) meeting in Chicago, he shared insights on how to lead:

Realize there's always a shortage of leaders, therefore plenty of opportunity to be one. This is true even in a society like ours, where the infrastructure is mature enough that A/E/C challenges are fewer than 20 years ago.

"A friend of mine has built a booming business—50 engineers—selling software for designing conical, reinforced concrete manhole structures," said Bannes. Creative, driving people make their own opportunities.

As much as anything, realize you lead by listening. "Many of us are rehearsing what we're going to say when we should be listening," Bannes noted. People who do that learn less than those whose brain is in the "receive" mode. And by listening they honor those around them.

Stand for the bottom line, and more. "We need to reemphasize engineers' code of ethics," said Bannes, who pointed out that a two-year-old Missouri law mandates that even politicians must now have a code of ethics. "We ought to write our own [code meaning more to us than existing ones seem to] or they'll impose one."

He recalled that when his former construction company was bidding a job, "a young engineer came to me and said, 'One of our people is exchanging notes with other contractors as to who's going to bid what, prior to bid opening.' We soon got back on track. When word gets around that you're honest, it's a wonderful marketing tool."

Share what you know with the next generation. "Early in my career as a structural engineer, I worked for an engineer who had trouble training me. He was only 10 years from retirement, and not open," Bannes said.

In contrast, one of his own hiring criteria is, "Find people smarter than I." And he tries to teach them everything he knows.

Infuse the organization with a sense of urgency. "It bugs me when I have to call someone eight times to get an answer," Bannes said. "We return our calls the same day," after the first call.

Find ways to empower others; you'll find that empowers you. Bannes asked, "When did you first feel oneness with moral authority? That you could do something well? That you had matured, become self-empowered, knew you could make it on your own?"

He said that "in partnering [one-on-one as well as in construction projects], we share this empowerment. Partnering is simple—be for someone else the kind of person you try to be for yourself.

"Do you have a moral compass?" Bannes asked the audience. He elaborated by relating the following incident: "I was teaching ethics to 200 senior engineers. One day at break a tall, bright foreign-born engineer came up to me and said, 'I hear what you're saying, but I don't understand it.

"'I was born in Calcutta, and don't know my natural parents. My first recollection is of the streets...the gang took care of me. A British couple adopted me when I was six, and sent me to the best schools. I know I have a good mind, but I have no formal religion, have never been to church. I'll be successful, and am sure I won't kill anyone.' But he sensed something was missing.

"Five years ago he wrote me that two years before he had married, and now had two kids. And he wrote, 'Thank you, I found my compass.'" Bannes had taught him about ethical behavior, had tried to personify the ethical person. He had become a role model.

Another example came up during the APM presentation. A member of Bannes' audience noted, "Minority contractors are accustomed to contracts sealed by word of

(Continued)

mouth, by a handshake. They're ill-prepared for bigger jobs [that come as a result of MBE/WBE setasides and quotas and goals]. How can we help them?"

Bannes recalled, "We were CM on a $3 million public medical center job. Two-thirds of the dollar volume went to minority contractors, and 90% of on-site workers were women and minorities. The job came in on time and under budget.

"How'd we do it? We teamed. There was a good-sized brick contract, and it involved three [minority] mason contractors [on the job] who'd never had a job this size. We teamed them with a good, large contractor who supplied the lead mason, and the three minority subs did the other work. Every Wednesday night we taught classes on such things as how to bid, how to estimate. We capped it off with a diploma.

"You hear of problems with minority contracts," Bannes said. "Many people give marginally qualified contractors enough rope to hang themselves, and then say, 'I told you so.' We must help them crawl, then walk. Seven or eight contractors on that job have now been mainstreamed."

More information. Lorry Bannes heads Bannes Consulting Group (St. Louis, 314-645-4665). The meeting sponsor, the Association for Project Managers, is in Chicago (312-664-2300).

Reprinted with permission from *Principal's Report,* Institute of Management & Administration, New York.

It was a cold, windy afternoon in December 1971 when, in a small but neat mid-town office, a nun, a young contractor, a retired postal administrator, and an attorney wrote an agreement to plan and build a multimillion-dollar medical clinic in inner-city North St. Louis. Our goal was both to serve the city's health care needs and to increase minority participation in the construction industry.

"Partnering" couldn't yet be found in America's dictionaries. But approximately two years later a health clinic opened its doors to the North St. Louis community. Construction was completed on time, in budget, and in strict compliance with the scope and quality standards of the plans and specs. Outwardly, just another successful project. But almost 90 percent of on-site work was by minority workers, and 60 percent of contract, supplies, and services work was awarded to minority firms. Of course, this didn't "just happen."

Project Goals

Sister Dolores McGhee, D.C., started each project team meeting with a prayerful reflection. I found this moment of pause to be a "compass spinner" which helped all of us, at least momentarily, to focus on accomplishing our task. Later, in some meetings involving tough agendas and difficult issues with uncooperative participants, I saw her meditations disarm the enemy—just one of those unfair weapons a 105-lb nun uses effectively that couldn't possibly work for anyone else. It was much like the time during the course of the project when we detoured to a chapel before a weekly progress meeting. When I asked why, Sister simply smiled and said, "Because...this is the first time we...don't have any money to pay you...." Disarming, yes, but very effective. So much so that I suggested we return to the chapel because I was capable of much better prayer!

At our first meeting, she outlined in her direct and simple way the goals of constructing a health center utilizing, as much as possible, minority and local neighborhood workers and firms. This was to be in addition to the usual parameters of cost, time, quality, value, and safety. To maximize chances of successful completion, the team would focus on open communication, education, and training. Project goals were coauthored by representatives of the owner, architect/engineer, and construction manager.

Contractor Selection Process

This project delivery system included fast-track construction at the urban renewal location. During the design phase, local minority contracting firms were studied to determine their number, sizes, and services, as well as strengths and weaknesses. Once these data were compiled and tabulated, the minority contracting capabilities were compared to our project needs and a strategy was developed to maximize their participation.

In each work category where numerous minority contractors had capabilities equaling or exceeding project requirements, one or more competitive bid packages was created. Excavation, grading, hauling, equipment rental, and fencing, plus finish, material supply, and general condition service items were included.

Other items of work coincided with minority capabilities, but the scope was too large for any one contractor to handle. Here, partnerships were created and/or the contract scope was broken down into phases or smaller packages. Divisions such as masonry, carpentry, drywall, and finish items, as well as plumbing, sewers, and electric, were accommodated in this manner.

Finally, in CSI divisions of work where few or no qualified minority contracting firms were available, joint ventures or GC/sub partnerships were developed involving majority subs and small emerging minority firms. Finish work such as terrazzo, and mechanical work such as fire protection and HVAC were handled in this way.

Potential participation in this project was marketed and sold extensively in the minority community. Many presentations were conducted for minority contractor groups. Selling was based on opportunities to acquire work within their capabilities, to complete their contract commitment in an atmosphere of supportive assistance, and to grow during the process.

Many traditional standards of prequalification were used in our evaluations, including experience and expertise, time and cost management strengths, individual and corporate reputations, and readiness for such a project. Nontraditional measures of qualification and characteristics were also developed and used, such as determining compatible philosophies and corporate cultures, ethics, and value systems. But this unique project also demanded that we find a way to determine a firm's creative problem-solving abilities, dispute-resolution attitudes, its "win/win" conceptual thinking strategies, and its openness to addressing challenges jointly.

In short, we tried to make this project very special to all participants. In return we asked that they "give back" by contributing interest and enthusiasm, talent and energy, time and open-mindedness. Overall, we stressed a willingness to listen, learn, and experience a positive growth process.

This experience provided me with a 40-year career involving nearly 1000 construction projects, one of my most meaningful lessons in how to succeed in business. Often, as my Mom related to me in early childhood, specialness is hidden, and only those who keep on patiently looking will find it. Often, the "key" to project success is related directly to finding its special features worthy of all our efforts and energy. Even the pedestrian warehouse or simple foundation system involves some reasons to celebrate if we keep looking for them.

Often it is a wonderful building design, or perhaps a construction challenge, an opportunity to stretch the limits of your capabilities, perhaps demand that you develop new ones. Sometimes it's the site, the schedule, the delivery systems, the adjacent environment, but it's always there if we just keep looking. And, of course, it can be found in the people associated with the project.

This project had it all.

Scope and Conditions/Communication and Analysis

While 18th and Biddle Streets in North St. Louis City had a rich history, almost all that remained were old granite curbs and cobblestone streets. The inner-city neighborhood's buildings had been torn down.

One project goal was to create a safe environment, so the project included fencing and security services.

Envisioning security of another sort, we shared the project vision with neighbors, public housing residents, the church-based community, and local businesses. This was to be their clinic, so we held information sessions, job progress reports, and on-site walk-throughs.

A minority contractor development and assistance office was set up near the site, to provide project communication and technical support.

It quickly became a busy meeting place. Workers here learned quantity takeoff analysis, checking and evaluation services, as well as overall project scheduling, including detailed component phasing and individual trade scheduling requirements.

Plan and spec technical requirements and design intents were also clarified. This meant constant communication with owners' representatives, designers, and construction managers. The principal benefits of these communication efforts came, I believe, in the areas of general requirements and conditions. I had noticed over the years that young minority contractors usually performed very well when conditions on the job site permitted and encouraged good and timely work. So on this project we tried very hard to create that kind of work environment. The key was that the contractors were given a clear understanding of work requirements and conditions.

This continuing dialog among owners, planners, designers, manager, and contractors created a win/win attitude among team members that would continue to grow throughout the project.

During the preconstruction analysis and bidding period, we were not only preparing to build, but teaching attitudes toward design and construction. Team members created detailed plans of how to build—including means, methods, and equipment sharing. But this project forced attention on aspects that are often taken for granted. Workers also detailed tool and supply requirements, individual trade manpower leveling, and integrated time, place, and resource allocation phasing and scheduling.

Constructability reviews, alternate/option reviews, and what-if analyses led to value engineering studies, sparking changes in the bidding and contracting documents.

During this prebid phase of study we also introduced the concept of detailed quantitative risk analysis, which served the contractors by arming them with detailed scope and condition evaluations. This provided the basis for a "gut check" that served them well on this project and, we hoped, throughout their careers.

We suggested a complete, similar, more pessimistic estimate based on less optimistic unit pricing, sub and supply imperfections, time delays, etc. We suggested keeping all these possibilities within the range of previous experience, although not as optimistic as a competitive marketplace estimate. Once this second estimate was available, a comparative study was made and a fee-versus-risk analysis was done, underlining those very basic principles of entrepreneurial construction contracting.

During this process we also found ourselves in a mentor/coach role as we helped minority contractors understand their own businesses more fully. Simple awareness of the costs of doing business were enhanced by studying their histories, more specifically fee income (anticipated and actual), the size, nature, and volatility of main-office costs, general and administrative expenses, and owning and operating capital requirements, and how they can vary and be controlled.

We kept referring to this initial analysis and estimating phase as our first rehearsal and suggested to all contractors what a wonderful opportunity we all had to build the job first in our minds and on paper. (One successful minority contractor who shared in this project told me several years later that he did not have an estimating department, but a "building the job on paper" group. I think he got the message and that it worked for him.)

Contracting Phases

We called our contracting phases the second rehearsals. The process of converting the bidding commitments into legally binding contract agreements can and should be another opportunity to do a practice run. In many projects, actual contract forms of agreement (sub, supplier, services) are predetermined and communicated during the bidding period. Such was the case on this project, and we implemented a long and very specific "plain talk" contract, spelling out terms of cost, time, quality, value, safety, and people performance.

Typical trade contracts (direct with owner) included dollar values, trade schedules, and crew size commitments for various phases of their work. Our cost breakdown was very detailed and broken down into labor, material, and equipment for each subcontract. Similarly, each subcontract had its own schedule. Thus, during construction it was very easy to check subcontractor progress, and to agree on future progress payments.

Fitting together hundreds of trade contracts, material supply orders, and service vendor agreements is often like putting together a jigsaw puzzle: One is trying to assure wholeness and completeness without gaps or overlaps. Interrelationships, shared resources, and mutual responsibilities were identified.

Temporary facilities and resources to be provided were documented, discussed, and dealt with—such as inclement weather, access, parking, snow removal, water pumping, cleanup, and debris disposal.

Technical or operational assistance to be provided by coaches/mentors or joint-venture partners was spelled out in all agreements. If contract administration/management weakness needed to be shored up, then such services were discussed and provided. Provisions for financial management assistance, including cash flow and progress payment billing, were arranged when requested or necessary and were spelled out in contracts.

Insurance, bonding, and banking representatives were kept informed of our plans, contracts, and risk management procedures, which helped to solidify and improve these contract relationships.

As we concluded each contract agreement and signing, we went over a simple but effective second-rehearsal "gut check." We asked if each party to the agreement had a clear understanding and working knowledge of the job requirements. Schedule commitments, site conditions, shared workplace adjacencies and co-responsibilities, anticipated profits, risks taken, and expected support were all discussed.

Affirmative handshakes at contract signing always remind me of previous times in our industry when a handshake alone served as a contract. In this case the handshake was like the one you get from the principal or dean at graduation. Now our graduates were going to work—constructing the building. The detailed preconstruction analysis and planning sessions had been our school.

Construction Preplanning

We identified the construction preplanning phase as our dress rehearsal—our third opportunity to practice building before we actually did it. All the actors had been cast, and a prestart meeting was held with each contractor several weeks before his work was scheduled to start. Contract requirements were reviewed in detail with key on-site and office personnel. Job site/building shell walk-throughs were conducted to familiarize each trade with the actual project conditions. A review of all submittals and approvals was completed. The status of all on-site materials, anticipated deliveries, unloading and storage requirements, as well as safety and protection of the same were checked. Manpower requirements, crew makeup and sizes were discussed, while initial mobilization and layout were examined as required.

On-site project administrative matters were reviewed with specific instructions concerning meetings, work hours, parking, safety procedures, etc., and were also included in this final check.

Office procedures for billing, change orders, and delay claims were also reviewed.

Conclusion

Having successfully built this job on paper three times, the actual construction was truly the easiest and most fun part. The partnering concepts developed and communicated during the planning phases promoted excellent communication.

The normal or typical construction management services were of course provided, supplemented by special services created for this project.

I started teaching weekly evening classes during the construction period (and beyond) at an adjacent office building and later in the health clinic itself, as construction progress permitted. The classes were open to all workmen and minority contractors, without charge. In fact, anyone interested, whether participating on this job or not, was welcome.

We viewed the project as an ongoing case study that provided a full range of real-life "look and see" examples of all subject matter presented. Team teaching, coaching, and mentoring techniques were utilized to bring in specialty resources.

Subject matter was suggested by the participants and included estimating and planning, leadership and management, co-ownership and operating requirements, accounting and financial management, and outside consultant support.

This sharing experience was very well received, and continued for years after the project.

A special dispute resolution program was created, which we called our "nip it in the bud" philosophy. One of the standard agenda items in each weekly construction meeting was to question any and all potential changes or problem areas. The mental set was revised from traditional avoidance to a welcome, "Let's hear both sides of the problem." The idea was that together we could solve any problem—provided we had sufficient time. Early detection was the key element. We had some early successes, and the approach became contagious among the team members.

Since quality and performance problems were anticipated, we also implemented an extensive sample, prototype, and mockup program for almost every item in the project. Whenever possible, we utilized these techniques to assure all parties, including the contractors, that their work and methods met the approval of architects, engineers, consultants, owners, and using agencies.

Looking back on the construction phase, it was truly the most enjoyable and rewarding aspect. We saw plans fulfilled, goals and objectives not only met but in many cases exceeded, simply because we worked in an environment of empowerment and optimism, which workers coauthored, and in which they grew.

Postconstruction

We stressed from the beginning that on-site completion was not the whole story—that punchlists, loose ends, and agreement finalization were also important. If these very real contract obligations are neglected, they can negate all the achievements, contributions, and services during construction.

Throughout construction, we developed a no-punchlist/no-callbacks project goal which involved inspections, reviews, corrections, and completions as the work progressed rather than after it was substantially complete. This involved partnering with designers and inspectors, and complemented our prototype/mockup program. It was much easier and more effective to accomplish punchlist work with the trade mechanics while they were on the job, rather than after they had left.

We showed how good, consistent record keeping could accomplish much in terms of the final contract requirements. Items such as as-built drawings, warranties, guarantees, operating and maintenance instructions, and the many lists and submittals can be, and were, assembled before construction was complete.

Probably the single most important and valuable postconstruction element of our partnering plan was our postmortem (exit interview) analysis. In concert with our final goal—to make this a positive growth experience—we conducted detailed, "How did we do?" evaluations with each team member.

Final, actual-versus-planned information was openly shared. Perceived (and actual) strengths and weaknesses of the participants and the relationships were discussed. As always, the meeting was conducted in a caring atmosphere of trust and with the intent of fostering growth. We discussed what events and procedures worked, which did not or could be improved. Nothing was sacred or spared our dialog—from nonproductive costs to personalities and leadership/management styles. We simply wanted to know how well (or how poorly) we had met our goals, and how we could improve.

Marketing the Next Job

Individual letters of appreciation/commendation/recommendation were prepared for each trade contractor, outlining specific planned, managed, and achieved goals. We encouraged their use in marketing/selling plans and/or presentation packages for bid list inclusions requests, RFQ/RFP, and/or cold-call leave-behinds. We suggested contractors make it standard operating procedure to request such letters of recommendation at the completion of every successful contract.

Summary

Over a career of nearly 40 years I have been involved in more than 1000 construction projects. The Courtney Health Center occupies a very special place in my memories, not so much because the team met and in fact exceeded most goals, but rather because of the caring relationships of good people. Team building and win/win partnering can and do work.

Epilogue

During the summer of 1972, Lorry T. Bannes, P.E., successfully defended himself and his company from claims of "reverse discrimination." It was alleged that they illegally took work from majority contractors by creating and implementing departures from traditionally accepted methods of involving minorities in the construction industry.

Not long after that, Mr. Bannes was presented with a National "Build America" Award by the Associated General Contractors of America for this project, including its educational opportunities provided to effectively help minority contractors. This project was the keystone achievement in a long list of experiences of this type.

Acknowledgments

I would like to thank: Sister McGhee (Chairperson of the Board, JCHS, Inc.); City of St. Louis (owner); Arthur Kennedy (deceased); Judy McKittrick (CDA Fund Manager); Walter Wren (Minority Contractor Consultant); Bernie Duda, P.E. (Project Manager, Bannes-Shaughnessy, Inc.); Paul Saunders (Executive Director Health Center); Wilbur Thomas (Associate Director Health Center); Godfrey Padberg (Dept. of Health & Hospitals, City of St. Louis); Fred McKissock; Margaret Wilson (attorney, NAACP/National President); Eddie (Johnson) Hussan, Minority Contractor (Clerk of the Works); and John Mickelletto, AIA, Architect-in-Charge (Charles Fleming, Architects).

Chapter

17

How the New Model Subcontract Was Created

Gerry Graff

Look for partnering examples other than single-project ones, and one of the more notable cases you will find is the new model building subcontract. How the Associated General Contractors of America (AGC), American Subcontractors Association (ASA), and Associated Specialty Contractors (ASC) created it is told in this chapter by 1994 ASA President Gerry Graff of Graff Flooring (Albuquerque, New Mexico).

From the start, the effort to produce it was joint, in contrast to the earlier AGC Document 600 model general contractor–subcontractor contract. The AGC prepared that contract by itself, and then asked the subcontractor community to sign it. This never happened, and the document is no longer distributed.

This time, a neutral, expert construction attorney, William Ferguson, then of the Boston firm Gadsby & Hannah, was retained to provide guidance. He was jointly selected by the association representatives following an elaborate process. First a request for proposals was issued inviting construction attorneys to submit position papers. The authors of what the group thought were the best papers were then interviewed, and Ferguson was chosen.

The team wrote its own marching orders before it started to draft the document. They outlined four guiding principles: (1) allocate risk fairly; (2) treat both parties equally; (3) recognize current realities of the construction industry; and (4) confine contract language to issues that are the norm and not the exception.

The document was not rushed into print. Two years were spent writing it and in legal review. The draft was then submitted to the sponsor societies, whose boards were told they could only vote it up or down. They could not "pick it to death" as such groups have been known to do. The societies approved it in spring 1994, and it was published by all three sponsoring groups that summer.

A New Contract

"Hell freezes over" read the page-one headline in the May [1994] issue of the newspaper of the American Subcontractors Association (ASA). The reason: Although subs and GCs have fought over contract language for years, ASA, the Associated General Contractors of America (AGC), and the Associated Specialty Contractors (ASC) had finally confounded their skeptics and arrived at an agreement.

Although most of the early interest focused on details of the contract, AGC's Grant Hesser and ASA's Collette Nelson both told CBMR that the document is most important as a symbol: GCs and subs have shown it's possible for them to act together.

That said, here are some of the key provisions of interest to subcontractors:

1. Pay-if-paid is out. "There was consensus early on that shifting the risk of the owner's nonpayment to subcontractors is wrong," commented William Ferguson, then of Gadsby & Hannah (Boston), who was retained by the contractor groups as their neutral legal expert.

2. Subcontractors should be involved in preparing the project schedule. After all, both parties will be bound by it. And the likelihood of completing the job on schedule is increased when all parties are dedicated to the effort.

3. The subcontractor has the right to receive copies of all documents to which it is to be bound, the GC's payment and performance bonds, and any other information the GC has obtained with respect to the owner's ability to pay.

4. The subcontractor generally receives payment within seven days after the owner's progress payment for the sub work is received.

Among the many GC concerns addressed:

1. The sub is liable if it fails to: provide the required bonding, report inconsistencies and omissions, visit the site prior to signing the subcontract, properly store materials supplied by others, or honor the warranty.

2. Relatively short periods of time are provided for the sub to react and/or give notice of inconsistencies and omissions, personal injury, and/or damage to the work, claims relating to the owner, and claims relating to the GC.

Will the contract be widely accepted? Attorney Ferguson predicts widespread adoption, if for no other reason than it's what the two parties—GCs and subs—wanted.

But ASA executive director Colette Nelson is less certain. Although she was bombarded by phone calls between the time of the document's completion in mid-April and its scheduled release at the end of June, she says it's possible the major pre-existing GC–sub model contract document, AIA A401, will continue to be the most widely used.

One potential problem is the length of the new document. At 40 pages it may be too ponderous a tome for many subcontracts. Ferguson predicts it will be widely used in all but small GC–sub contracts—those of less than $100,000. And Nelson says the length is not intimidating. "For a given job, you look only at what's been changed." You read only the underlined words (those the other party proposes adding) and those crossed out (to be deleted).

(Continued)

> For the time being, no electronic version will be available. Says Nelson, "Electronic documents are relatively new. It was decided to study and learn from what happens with AIA's."
>
> To order the *Standard Form Building Construction Subcontract,* contact one of the three sponsor groups: AGC (Washington, D.C., 202-393-2040), ASA (Alexandria, Va., 703-684-3450), or ASC (Bethesda, Md., 301-657-3110).

Reprinted from *Contractor's Business Management Report,* 1994, Institute of Management & Administration, New York.

It has been common to think of the general contractor/subcontractor relationship as employer/employee, but another analogy is that of a partnership of teammates. Here's the story of how, using a bit more of the latter analogy and less of the former, in 1994 several associations of general contractors and subcontractors crafted—and approved—a new model subcontract governing their relationship.

Partnering among contractor associations is not new. The three sponsoring groups have been working together for 20 years, and among the fruits of that cooperation has been a document. *Guidelines for a Successful Construction Project,* which in summary says:

1. The primary objective of every general contractor and subcontractor is to successfully deliver to the owner the specified project on time, at contract price, and achieve a reasonable profit. No general contractor can deliver a project successfully without the cooperation of competent subcontractors. And no subcontractor can perform its work successfully without a corresponding measure of cooperation and leadership from a competent GC. They need each other.

2. Both seek a business relationship on which they can depend. Usually, each wants to continue to do business with the other on future projects as well as those at hand. Skill, integrity, fairness, and responsibility will make the contractual relationship possible, profitable, and pleasant. Both know the keys to a successful relationship.

3. A written contract between the two parties is common. Experienced contractors know that no architect/engineer can prepare a perfect set of documents, and that no GC or sub can perform perfectly. Likewise, contract documents cannot detail every industry practice, or anticipate every crisis or situation that will arise at the site. Mistakes will occur, and miscommunications will arise. Errors and omissions, when discovered and made known in timely fashion, can generally be overcome with minimal damage to the parties. Mistakes that are admitted and corrected immediately are the least costly. Most GCs and subs solve their problems without resorting to litigation.

4. Each GC and sub on a project should be regarded with equal respect. Each is an expert in its own field. Ethical conduct regardless of contractual "rights" or ability to make another suffer is essential

for harmony. The Golden Rule—"Do unto others as you would have others do unto you"—is still the best guideline in dealing with other individuals and businesses.

How the Model Subcontract Was Written

It was in this spirit that Paul Emerick of AGC and Charles ("Rusty") Griffiths of ASC became encouraged to attempt to write a building subcontract. They felt strongly that with the renewed spirit of the industry—trying to settle its differences without going to lawyers or court—they should try again. Earlier efforts, beginning in the mid-1970s and ending in 1983, had ended in failure.

To that end, in November 1990 the General Premise of the subcontract was written. The objective would be a document which:

1. Allocated each risk to the party best able to control it
2. Provided contract terms based on equality of the parties
3. Was fair

Willingness to tackle something where earlier efforts had failed was slow to come. The next milestone came in November 1991. After encouragement, and the setting of some ground rules, the presidents of the three groups—Marvin Black of the AGC, David Miller of the ASA, and Theodore Brodie of the ASC—signed a memorandum of understanding to proceed with the effort. The stated goal was to develop, for joint publication, a standard-form building subcontract which would define for the most common building construction project situations the rights, responsibilities, and relationships between a GC and subcontractor.

The memorandum provided for appointment of two representatives from each association (no one of whom had previously negotiated on the subject) and one staff representative (see sidebar on p. 263). It was agreed that for the effort to succeed, a fresh start would be important; any previous "baggage" should be eliminated.

There would be no publicity, it was agreed. Also, should the discussions be abandoned, the parties agreed not to discuss or publicize the reasons why. It was important that liaison between the groups continue, even if this effort failed.

At the subcontract writing team's first meeting, in January 1992, they devised their strategy. The most common approach in such efforts is to tackle the easier topics first, thus building rapport. That way, when they get to tougher areas, there is momentum and better understanding of others' views. The contractors, however, decided on the opposite approach.

They knew they had a monumental assignment: Despite previous attempts, a model subcontract document with which all three associations agreed had not been developed. Although the ASA and ASC endorsed the AIA's subcontract agreement, A401, the AGC did not. The AGC had its own subcontract—AGC 600—but the other two groups did not endorse it without significant revisions. In fact, the ASA and ASC published a commentary AGC Form 600, which recommended extensive revisions.

It was unclear whether the new group would succeed. While hoping for success, most members worried that their talks might be futile.

As a result, they decided to tackle the toughest issues first. Thus, if they could not agree on the larger issues, a minimum of time, energy, and funds would have been wasted.

Next, secret ballots were used to determine on which issues there was the most disagreement. These ballots led to the following order of discussion:

1. Payment and final payment (including retainage, the owner's ability to pay, stored materials, noncontracted services, late payment interest, and subcontractor payment failure)
2. Indemnification and additional insured clauses
3. Performance issues, including bonding and like securities
4. Flow-down clauses
5. Shop drawings and professional services
6. Warranties
7. Concealed conditions
8. Hazardous materials
9. Schedule of work
10. Correction of work
11. Safety
12. Scope of work

With this list compiled, the immensity of the project became more apparent. At the same time, the team felt if they could agree on the first three or four issues, then the probability of success would be much greater.

The committee agreed to work without an attorney until the next milestone, completion of the Agreement in Concept. This would describe the sense of the committee and the principles of the subcontract, but would not contain its language. The team did agree that the contract would be compatible with the most widely used owner–GC contract, AIA A201.

The group initially expected to write the Agreement in Concept in three one-day meetings over six months, but soon found that it would take more time and switched to two- and three-day meetings. The latter proved to be too much, both mentally and physically, so two-day meetings became the norm. Even with the far heavier meeting schedule, the team did not complete the Agreement in Concept until August 1992.

All negotiations were kept confidential. So tight-lipped were the members that after a few months the group became known within the associations as the "Stealth Committee."

Although it had been agreed to leave personal experiences out of the discussions, intense personal experiences did come to the surface. These in fact helped shed additional light on the issues. None of the six negotiators was inexperienced at construction, yet each had something to learn from the others.

The group also found that the initial list of tough issues was off the mark. Some issues that were fairly low on the list proved as difficult to resolve as others higher up.

On the other hand, the group found that it had another advantage: Normally, a GC and a sub negotiate a subcontract after the latter has given a low bid on the project to the GC. How badly does the GC need the subcontractor (by how much was his bid low)? Conversely, how badly does the sub need the GC (how much does the sub need the work)? In this setting, other issues came into play. It was good to have those barriers removed from our discussions.

Retaining an Attorney

The next step was to hire an attorney to develop the subcontract language. Each association submitted names of law firms they felt could approach the task without bias. This list was narrowed to two firms, whose representatives were interviewed in December 1992. The unanimous choice was William Ferguson of Gadsby & Hannah (Boston).

The plan was for Bill to take the Agreement in Concept and draft the document. Once he had done this, the team felt that one or two more meetings might be needed to review it.

The Stealth Committee failed to realize that in many cases, in attempting to reach Agreement in Concept, issues had been left too broad, and open to a variety of interpretations. When Bill asked questions as to what the group meant by a passage, he got conflicting answers, even though the members had believed they understood exactly what the Agreement in Concept meant.

Further, as the language emerged, it became obvious that they had not explored all the important issues, or that revisions were needed to provide balance. Again, there were more meetings.

As the contract's language emerged, something very good, but not originally expected, was discovered. Since all the voting members of the Stealth Committee were contractors, the contract language that emerged was plain and clear. Many GCs and subs could understand most of it. It was refreshing to see wording that one did not have to take to an attorney for translation.

After the document was completed in December 1993, the next step was to obtain the final endorsement of the three associations' boards. The negotiators had recognized the political nature of their respective boards—an ASA board member, for instance, was not likely to put GC concerns first when voting on the document.

It was agreed the three boards would be told they could only vote the document up or down; they would not be allowed to pick it to death, as legislative bodies have been known to do.

By April 1994 all three groups' boards had approved the subcontract. It was published July 1, 1994.

The new *AGC/ASA/ASC Standard Form Building Construction Subcontract,* like its predecessor *Guidelines for a Successful Construction Project,* represents the best in association partnering. The local affiliates of the three associations, and other national and local associations, would do well to emulate their efforts. The example the contractor groups set will do much to make partnering a reality throughout the industry.

Key Provisions of the Model Subcontract

(a) PROVISIONS THAT REDUCE RISK TO THE SUBCONTRACTOR

- The subcontractor has the right to receive copies of:
    (1) All documents to which it is bound
    (2) The general contractor's payment and performance bonds
    (3) Any information the general contractor has obtained with respect to the owner's ability to pay
    (4) The Builder's Risk Insurance policy
    (5) All contract provisions pertaining to claims by the contractor against the owner
    (6) The most recent pay request submitted to the owner for subcontract work performed

- Payment provisions
    (1) The subcontractor generally receives payment within seven days after payment is received from the owner for subcontract work properly performed.
    (2) If for any reason not the fault of the subcontractor, the owner fails to pay the GC, the GC will pay the subcontractor within a "reasonable time." If subcontractor action is necessary to preserve lien or surety bond rights prior to the expiration of a "reasonable time," the subcontractor may take such action.
    (3) If the subcontractor does not get paid in a timely manner, it is entitled to receive interest. If lack of payment is caused by owner failure to pay

(Continued)

for a reason not the fault of the subcontractor, then the interest due the subcontractor is its pro rata share of the interest paid to the GC by the owner.

(4) The subcontractor has the right to stop work and eventually terminate the subcontract if it does not receive payment for work properly performed. Provisions are in place for cost recovery, should the subcontractor take these actions.

(5) The subcontractor will get written notice when its application for payment is disapproved, along with the reason for the disapproval.

(6) The amount retained from the subcontractor is equal to the amount retained by the owner. If the owner specifies zero retainage, then there will be no retainage on the subcontractor. However, if no retainage is specified by the owner, then the retainage will be negotiated.

(7) The general contractor cannot hold monies under this subcontract for work being done under a different subcontract for a different project.

(8) Under no circumstances can the general contractor require the subcontractor to sign an unconditional waiver of its lien rights prior to receiving payment.

(9) The subcontractor is reimbursed for its payment and/or performance bonds with its first pay estimate. There is no retainage on payment bonds.

(10) The subcontractor may bill for executed change directives, or portions thereof which are not in dispute.

- The general contractor cannot request a subcontractor to provide payment and/or performance bonds subsequent to signing of the Subcontract Agreement.

- The general contractor is required to develop the Schedule of Work substantially based on subcontractor input. If the schedule is changed drastically, it is considered a change in the scope of the Subcontract, and may potentially result in a change in the Subcontract Cost and/or Time.

- The general contractor has an affirmative duty to provide suitable areas of storage on the site, and generally will reimburse the subcontractor if it requires the subcontractor to relocate the materials.

- A subcontractor cannot be obligated to pay a general contractor's OSHA fines unless, and only to the extent that, the fines are a direct result of the subcontractor's actions. In any event, a subcontractor shall not be obligated to pay that part of the general contractor's fine that is the result of repeat violations.

- The general contractor must inform the subcontractor if Builder's Risk Insurance is not in place and, in such case, reimburse the subcontractor for purchasing the coverage.

- Generally, if the subcontractor is delayed for no fault of its own, it is entitled to a time extension.

- If liquidated damages are assessed by the owner, the general contractor can only charge a subcontractor in proportion to its responsibility. That is, the general contractor cannot charge multiple subcontractors the entire amount.

(Continued)

(b) PROVISIONS THAT REDUCE RISK TO THE GENERAL CONTRACTOR

- The subcontractor is liable and there are specific remedies available to the general contractor if the subcontractor:
    (1) Fails to provide the required bonding
    (2) Fails to report inconsistencies and omissions
    (3) Fails to visit the site prior to signing the subcontract
    (4) Fails to properly store materials furnished by others
    (5) Fails to honor the warranty
    (6) Covers work which the general contractor has requested be left uncovered
    (7) Covers defective work
    (8) Does not correct rejected Subcontract work
    (9) Does not clean up after itself
    (10) Fails to perform the Subcontract work in a safe and reasonable manner to protect persons or property
    (11) Fails to comply with federal, state, and local laws, ordinances, and regulations or knowingly violates a patent
    (12) Fails to supply enough properly skilled workers, proper materials, or maintain the Schedule of Work
    (13) Fails to make appropriate payments to others to which it is obligated

- There are relatively short periods of time for the subcontractor to react and/or give notice of:
    (1) Inconsistencies and omissions
    (2) Personal injury and/or damage to the work
    (3) Claims relating to the owner
    (4) Claims relating to the contractor
    (5) The subcontractor's Schedule of Values
    (6) The consequences of general contractor suspension for convenience

- The general contractor has adequate protections in the event the subcontractor is not performing or is being financially irresponsible with regard to the project.

- The general contractor has the right to change the Schedule of Work as required to get the job done.

- The general contractor has a say in the subcontractor's subcontractors, material suppliers, etc.

- The general contractor may order the subcontractor in writing to perform incidental work without a change to the Subcontract cost or time.

- If the owner assesses liquidated damages, the general contractor may charge the responsible subcontractor(s) as well as recover from the subcontractor the actual damages sustained by the general contractor.

(c) PROVISIONS THAT REDUCE RISK TO BOTH THE GENERAL CONTRACTOR AND SUBCONTRACTOR

- Additional services provided by either party, except in the case of a safety hazard, require notice prior to providing the additional service.

(Continued)

- Prevention of accidents is everyone's responsibility; work is stopped when unsafe conditions or substances are encountered.
- Both parties are bound by the Schedule of Work.
- The Subcontract is governed by the law of the state in which the project is located.
- In the case of a dispute, the prevailing party is entitled to recover reasonable attorneys' fees, costs, charges, and expenses.
- Proper insurance must be maintained by all parties.
- Reasonable provisions exist for changes in the Scope of the Work and for dispute resolution.

Negotiators Representing the Three Associations

Representing the Associated General Contractors of America:

Grant Hesser,
Maescher Industries,
Cincinnati, Ohio

Robert Martin
F. H. Martin Construction,
St. Clair Shores, Michigan

Cheryl Terio,
Associated General Contractors,
Washington, D.C.

Representing the American Subcontractors Association:

Gerry Graff,
Graff Flooring Contractors, Inc.,
Albuquerque, New Mexico

Peter Hess,
Albany Steel,
Albany, New York

E. Colette Nelson,
American Subcontractors Association,
Alexandria, Virginia

Representing the Associated Specialty Contractors:

Charles ("Rusty") Griffiths, Jr.,
Binghamton Slag Roofing Co.,
Binghamton, New York

Ernest Menold,
Ernest D. Menold, Inc.,
Lester, Pennsylvania

Daniel Walter,
Associated Specialty Contractors,
Bethesda, Maryland

Part 4

Tomorrow

Chapter

18

Can Partnering Transform Design Practice?

Kyle V. Davy

In this book partnering is defined broadly. Look for the most dramatic construction stories, the biggest breakthroughs in terms of cost reduction or by other measures, and you will find that in every case, exceptional teamwork or partnering was involved. Kyle Davy, an A/E management consultant (Berkeley, California), cites the partnering between R. J. Reynolds Tobacco Co. (Winston-Salem, North Carolina) and its contractors.

They did away with bidding and virtually eliminated agreements. Instead of sophisticated management control systems designed to manage large projects, they divided big ones into small work packages ranging from 7000 to 12,000 ft^2 of building space each. These smaller packages could be easily executed by small teams. Reynolds awards work packages to "partners" on a time and materials basis. Long-term partnering relationships allow teams to move up learning curves toward improved schedule and cost performance. They have cut units costs and cycle times over 50 percent.

Partnering has changed—and improved—the way projects are designed and constructed. Partnering projects consistently show better on-time and on-budget performance, enhanced safety, reduced claims activity, and increased satisfaction by project participants, especially owners.

As impressive as these direct project benefits are, the biggest benefit from partnering may come from the transformation it sparks in the way design firms practice. This transformation focuses on three major strategies:

- Building *real* teams throughout a firm to achieve new levels of performance
- Creating a learning organization by investing in the training and development of staff
- Breaking the rules to fundamentally redesign the way design firms do their work

What Is Partnering?

Partnering is a "covenant of good faith"[1] among the project parties. They agree to operate on the basis of shared goals for the project rather than individual organizational agendas.

For the duration of the project, all parties commit to a relationship based on trust and open communication. The partnering process builds this new relationship in three ways, through team building, collaborative problem solving, and goal setting.

- *Team building* allows people to get to know each other and appreciate their differences. Participants learn each others' roles, the risks and concerns they feel, and what they need to be successful. They also learn the importance of listening and improve their communication skills. Jointly, they identify obstacles to building a successful relationship and ways of overcoming them.
- *Collaborative problem solving* includes early identification and work on critical project issues. Participants agree to address ongoing problems in an open, timely fashion. They learn group problem-solving, decision-making, and conflict resolution skills and tools. The team also designs and implements a system for resolving future issues and disputes.
- *Goal-setting* activities include the establishment of shared goals and objectives for the project. A "partnering charter" is signed by each of the participants, symbolizing their commitment to the process and to each other. A joint evaluation process is adopted to allow the team to monitor its own performance and to assess the effectiveness of partnering activities and attitudes.

Much of this work is done in a "partnering workshop" at the start of the project. The workshop brings all of the major stakeholders for the project together in a retreat setting, where they can concentrate on the partnering process. Outside facilitators often help organize the workshop, provide training (in communication, problem-solving, and conflict resolution skills), and guide the group through the team-building, problem-solving, and goal-setting efforts.

Periodic partnering meetings, ongoing communication and problem-solving activities, and joint evaluation efforts provide continuity for the partnering effort during the remainder of the project.

The Benefits of Partnering: Changing Projects and People

Partnering works, for the entire project team. Owners, contractors, and design firms all attest to the benefits.

The U.S. Army Corps of Engineers has used partnering on over 100 projects to date. Results have exceeded their expectations over 90 percent of the time. Schedules shorten and costs fall. Value engineering opportunities are more likely to be identified and implemented. And there has been no litigation resulting from any of these projects. In 1990, the Corps of Engineers enthusiastically adopted partnering as an organization-wide initiative.

Contractors like partnering too. Builders report an increased ability to "earn their as-bid profit" on projects using partnering through the equitable resolution of issues, mitigating the need to "make it up on change orders." This type of experience by MCI Construction Company is described in Chap. 4.

The resulting drop in claims activity has been a big money saver. The president of one large contracting organization, speaking at an industry conference in 1992, noted that partnering had cut his company's legal fees by two-thirds over a three-year period, a multimillion-dollar saving. McDevitt Street Bovis's experience, discussed in Chap. 15, shows the dramatic impact of this type of result.

Design professionals, originally slow to get into the act, are signing on. In their new handbook on partnering, the American Institute of Architects and the American Consulting Engineers Council note that "The benefits are clear: Projects are completed on time, within budget, to high standards, and to the satisfaction of everyone."

Everyone reports dramatic reductions in paperwork and posturing. In commenting on the success of partnering for his project, one project manager noted that, "Normally, I'd have a stack of paperwork up to my shoulder in preparation for future claims activity at this point in the project. In fact, right now I only have a notebook about one inch thick."

In an environment of trust, open communication, and expedited problem solving, parties find that it's OK to take the risk of communicating orally to solve problems and make decisions. Written documentation of final decisions should still be made, but can come later.

As adversarial maneuvering and defensive hostility are replaced by joint problem solving and open communication, professional satisfaction increases. The project environment begins to "feel like it used to, years ago, before we all learned how to think and act like lawyers."

One facility's director tells the story of an individual who had just come off five years of work on several troubled, litigious projects. He had developed a reputation for being harsh and short tempered, and someone others didn't want to work with. During the course of his first partnering project, he changed into, or back into, an open, pleasant, productive team member.

Partnering as a Bridge to Total Quality Management

The origins of partnering can be traced to early attempts by large private companies to apply the lessons of continuous improvement to their business operations. In the early 1980s, companies such as Ford and Du Pont sent people to Japan to discover how Japanese companies had made such large strides in improving the quality of their goods and services. They found *kaizen,* or the practice of continuous improvement, and committed themselves to implementing total quality management (TQM) throughout their organizations.

As they experienced success implementing TQM in their core operations, they naturally began to apply the same principles to all areas of their operations, including their capital project delivery process. Du Pont teamed up with large engineering and construction companies such as Flour Daniel and MK Ferguson as "partners" in a new process which used TQM concepts and tools and emphasized team building, increased communication, shared risk, and collaborative problem solving within the context of a long-term customer/supplier relationship.

For their first formal *public* "partnering" project, the U.S. Army Corps of Engineers set out to capture some of the benefits that Du Pont and others were experiencing through their use of these long-term "partnerships." Could you foster similar teamwork, trust, and open communication on public construction projects?

Working with an outside consultant, the Corps designed a "partnering" process for the second phase of its Oliver Lock and Dam project along the Black Warrior-Tombigbee Waterway at Tuscaloosa, Alabama, that achieved these goals. Their process was built around basic principles and tools drawn from TQM, including team building, process improvement and collaborative problem solving, and measuring and monitoring progress toward shared goals. (See Chap. 2.)

From these humble origins, partnering has swept the country. In barely five years, partnering for public construction projects rose from experimental status to standard operating procedure. Hundreds of projects have been completed and hundreds more are under way.

With this success, partnering and TQM came full circle. Faced with both customer and management demands to implement TQM, facilities, design, and construction organizations that were trying to figure out how to do it are realizing that an answer can be found in partnering.

Partnering can serve as a bridge that organizations can use to implement TQM. It teaches tools and shapes attitudes that organizations need to implement TQM. It creates advocates and leaders who believe in the power of open communication, continuous improvement, and collaborative problem solving. It offers a starting place for TQM that is tangible, easily understood, and doable.

USAA Insurance was an early leader in implementing TQM in its insurance operations. Faced with a mandate from management to implement TQM within the facilities design and construction operations, the director of the department decided to start with partnering. The successful implementation of partnering on several major building projects provided the perspective he needed to chart a course for the overall implementation of TQM within the department.

Ultimately, partnering finds its proper place as one of a growing set of quality management tools that organizations can use as part of comprehensive TQM efforts. It is a hybrid application that combines a set of quality management tools and techniques in a way that is uniquely suited to the needs of the design and construction industry—a way that changes dramatically how construction projects are designed and built.

The impact of partnering as a quality management tool used directly on projects has been significant, not only for the short-term benefits derived on individual projects using the practice, but as a result of the systemic changes that begin to happen within the parent organizations of the team members as they absorb three primary lessons that come with the use of TQM in general and partnering in particular.

Transforming Practice: Three Lessons from Partnering

The first lesson looks at the "structure" of partnering and focuses on teams and teamwork. The second lesson looks at the "process" of partnering and sees learning and skill development. The third lesson sees partnering as a "symbol" of change, pointing the way out of the vicious cycle of declining trust and satisfaction that characterized the design and construction industry in the 1980s.

Lesson one: the wisdom of teams

Construction projects have traditionally been organized around project teams. Carefully documented organizational charts identify project team members and how they will communicate and make decisions. Agreements codify the relationships, roles, and responsibilities of the team members. Unfortunately, most of these are teams in name only; some management experts call them "pseudo-teams." "Pseudo-teams" publicly proclaim themselves as teams, even though the individuals involved have little interest in shaping a common purpose or set of performance goals. Trust is virtually nonexistent, and individuals act according to their own organization's agenda, guided by political and legal concerns that bear little relationship to performance considerations.

Partnering, in contrast, builds "real teams" that substantially outperform traditional project teams—often not a difficult task. "Real teams" commit themselves to a common purpose, performance goals, and approach for which they hold themselves mutually accountable. They are composed of a small number of people with complementary skills. Team building teaches them to trust each other, to communicate openly, and to solve problems quickly. Members do "real" work together and measure their success by assessing their collective work products.[2]

Discovering the "wisdom of teams" is a recurring theme in many organizations restructuring to meet the demands of today's economy. Traditional hierarchical and functional structures, stovepipes and chimneys, are being replaced with "real teams" that can deliver the performance essential to meeting escalating demands for responsiveness, speed, customization, and quality. Self-managing teams are used to construct projects or manufacture products, quality improvement teams provide bottom-up change, cross-functional teams implement changes to core processes, and top-management teams provide vision and leadership.

"Real teams" seldom just happen; they are built. Prior to partnering, few mechanisms existed for building real teams on design and construction projects. Consequently, it seldom happened. However, partnering has reversed this condition.

From the start, partnering was designed to build teams. The design of the original partnering process for the Mobile District of the Corps of Engineers centered on group dynamics and team building. The success of this partnering process as it has spread across the country has given many individuals and organizations their first taste of the performance potential of real teams.

Partnering not only exposes design firms to the benefits of real teams, it also gives them first hand experience in how to create them.

The attractive returns from team building reinforce the necessity of this change, not just for partnering, but throughout the design and construction industry.

Teams transform the project delivery process. Design firms are adopting team structures and team-building processes across their projects and throughout their organizations. Many firms find that the first step is the application of partnering at an early stage of the project delivery process. This means transforming the project initiation process. Design firms which once plunged into the programming and design of a project upon notice to proceed are now "going slow to go fast," spending time up front on team building with clients and subconsultants. They are finding that the benefits of open communication, collaborative problem solving, and joint goal setting pay substantial dividends in the form of reduced rework and expedited schedules.

As design firms find opportunities to apply their new knowledge of teams to all phases of their work, partnering and team building become part of the standard repertoire of skills that a design professional brings to projects.

This new emphasis on teams and teamwork comes full circle for partnering when it is applied to long-term customer–supplier (client–professional) relationships. Businesses are recognizing their interdependence and "codestiny" with their suppliers, including design consultants. They are demanding new long-term "partnerships" with small numbers of suppliers who can prove their ability to work with them in teams.[3] (See Anthony Costonis's discussion of multiproject strategic partnering in Chap. 8.)

Partnering prepares design firms to do this. Firms that have learned the basics of teams and teamwork through the partnering process have a competitive edge over firms that have not. They have learned the right skills, and they are also able to teach their clients how to build "real teams." These are the design firms that will not only survive but will prosper in the coming business environment. A parallel future is advocated by James Bradburn in Chap. 5, where he advocates reuniting designers and contractors during construction.

Teams and organizational structure. The performance potential of teamwork can also be harnessed inside the firm, just as clients have reorganized their organizational structures and processes around teams. Traditional design firm structures such as studios, profit centers, and functional matrixes dissolve into self-managing design teams, where results are assured not by management command and control, but by high-performance teamwork at the project level.

Firms can also build internal teams to accomplish a variety of nonproject tasks. The most common application will be in the form of quality improvement teams focused on solving internal problems and improving work processes. Top management in design firms will also use team-building skills to improve communication and trust in the pursuit of a shared vision and strategy for the firm's future.

Lesson two: a commitment to learning

The rate at which individuals and organizations learn may become the only sustainable competitive advantage, especially in knowledge intensive industries.[4]
ROY STATA, CHAIRMAN
ANALOG DEVICES

Organizational learning has moved from academic theory to the cutting edge of management practice. Corporate leaders such as Roy Stata at Analog Devices, Paul Allaire at Xerox, and Gordon Forward at Chapparal Steel have focused their strategic plans on enhancing organizational learning. Their actions express a growing belief that the most successful organizations will be those that are best at creating, acquiring, and transferring knowledge, and changing their behavior to reflect new knowledge and insights.

John Dewey noted that "[all] learning is a continual process of discovering insights, inventing new possibilities for action, producing the actions, and observing the consequences leading to insights." Organizational learning aggregates the education and self-development activities of individuals into shared knowledge and insights that can be used to create an organization's future. This "public" knowledge helps members describe and consider:

- How we do things.
- Why we do things the way we do.
- What's changing—clients, markets, the economy, etc.
- How we should respond to these changes. What new things should we be doing? What old things should we stop doing?

The creation of this "public" body of knowledge takes on a life of its own in "learning organizations." Initial learning efforts lead into a continually renewing cycle of knowledge generation, communication, and change, where organizational learning becomes a sustainable, organic process.

Thomas Jefferson said that "a democracy is only as strong as its public education." In the same way, the ongoing success of a design firm depends on the extent and richness of its public knowledge base, and the strength of the learning processes that continuously contribute to and reshape it.

Partnering as a learning process. Organizations are constantly being challenged to discover new management tools and methods to accelerate organizational learning, build consensus for change, and facilitate the change process. A wide range of activities and media are available to aid in this quest—books, articles, conferences, problem-solving networks, visits and visitors, experiments and pilots, research initiatives, training, benchmarking, planning, and strategic alliances, to name a few.

One of the most important forms of organizational learning is total quality management. Craig Barrett, executive vice president of Intel, observes: "We neither present nor champion our quality effort under the banner of 'This is Intel's total quality effort.' Our quality effort is designed to bring in those tools, techniques, and concepts that are useful to the corporation to improve our processes, our customer service, and our performance against our set of values. These values are the driving bedrock of learning...."[5]

Within the design and construction industry, partnering has emerged as another foundation of organizational learning. The themes of partnering—trust and open communication, joint problem solving and continuous improvement, management by fact, cross-organizational management, and a commitment to teamwork—contribute directly to organizational learning.

David Garvin identifies three overlapping stages in organizational learning. First, members are exposed to new ideas, expand their knowledge, and begin to think differently. Second, members begin to internalize new insights and alter their behavior. Third, members change their behavior, effecting measurable improvements in results.[6]

Each of these stages occurs in partnering. Participants are exposed to new ideas and knowledge. These new ideas are internalized and manifested in changed behavior by designers, builders, and owners, both during and after the workshop. The monitoring system established as part of the partnering process provides positive, tangible evidence of the impact of changed behavior.

The initial forum for much of this learning is the workshop, with the facilitator serving as teacher and role model. Teaching includes appreciation of individual differences in learning and decision-making styles, team skills, interpersonal communication techniques, problem-solving road maps and decision-making tools, and partnering principles such as "win/win" negotiation.

Educational theorists note that learning is not just "taking in information"; learning is active. New insights and possibilities need to be accompanied by opportunities to act and to observe the consequences. Partnering provides these opportunities for action within the workshop and during ongoing project work.

Project team members also learn from each other in ways that were not possible before partnering. Participants are able to expand their knowledge of the work processes and priorities of their counterparts in the design and construction process. On-the-job training kicks in as owners, builders, and designers lay down their cards and share "secrets" while solving problems and improving critical work processes together. In the past, these secrets were closely held, driven by fear that anything disclosed would be used against the party that disclosed it.

Learning also takes place as individuals bring new ideas and knowledge back to their parent organizations. They often return as champions not only of partnering, but of new ways of doing things in general. Once they experience the success of a partnering project, it is difficult for them to return to unattractive old ways of doing work. Partnering provides the proof they need that learning is an effective strategy for improving group performance. This strategy pays off for projects, for people, and for firms.

From partnering to the "learning organization." Beyond creating champions for change, partnering prepares firms to become "learning organizations." David Garvin identifies five key skills that are the building blocks of learning organizations. Partnering lays a foundation for each of these building blocks, as shown in Fig. 18.1.

Beyond skill building, an investment in partnering also serves as a symbol of an organization's belief in the importance of learning. The organization's investment in new skills and knowledge acquired through the partnering process demonstrates its commitment to learning. It "walks the talk." Literally millions of hours of training have been paid for and provided to industry participants in the last five years within the context of partnering workshops. As this investment pays off, firms are more willing to put more money and effort into in-firm training.

Partnering and organizational learning, just like TQM and organizational learning, are inextricably intertwined. Partnering embodies a commitment to learning. Learning enables partnering. As Roy Stata notes, "organizations must learn to learn," to be successful in the future. Partnering begins this process for design firms.

Lesson three: breaking the rules

After a decade of total quality management and incremental change efforts, organizations are realizing that improvement of existing processes is no longer enough. The times demand a "fundamental rethinking and radical redesign of business processes to achieve dramatic improvements in critical contemporary measures of perfor-

> Building Block #1: Systematic problem solving
>
>> Partnering teaches new problem-solving, process-improvement, and decision-making skills. These skills are taught in the workshop and learned "on the job" as they are applied to project situations.
>
> Building Block #2: Experimentation with new approaches
>
>> Partnering introduces new ideas and knowledge drawn from continuous improvement and organizational development, such as team building and "win/win" negotiating.
>
> Building Block #3: Learning from their own experiences and past history
>
>> Partnering emphasizes data gathering, both for problem-solving efforts and for the monitoring process.
>
> Building Block #4: Learning from the experiences and best practices of others
>
>> Participants receive on-the-job training from their counterparts on the project. For example, understanding how contractors really handle submittals can lead to insights that design firms can use to improve their own internal processes. Also, the potential exists to "benchmark" outside the project as a means of setting goals and discovering solutions that might not have been obvious to project team members.
>
> Building Block #5: Transferring knowledge throughout the organization, quickly and efficiently
>
>> Partnering fosters open communication and trust, enabling rapid dissemination of information and identification of problems throughout both the project team and the parent organizations.

Figure 18.1 The five building blocks of learning organizations.

mance, such as cost, quality, service and speed."[7] This includes breaking the rules that limit an organization's ability to conduct its work.

Partnering is just this type of radical change—so radical that, to implement it, organizations leapt over two decades of finely crafted contracts embodying "risk management" prescriptions from attorneys and insurance companies. Instead of reforming or rewriting their agreements, parties simply committed to implementing partnering.

Having vaulted over the lawyers and insurance companies, partnering proceeded to break many of the rules that had guided the construction process for the previous 25 years. For examples of this, see Fig. 18.2.

By invoking a new set of rules, partnering begins the process of reengineering the construction phase of the project delivery process, and sets the stage for more fundamental change.

R. J. Reynolds Tobacco Company broke many rules in transforming its facilities design and construction process. Using a "just-in-time" concept, Reynolds did away with bidding and virtually eliminated agreements. Instead of sophisticated management control systems designed to manage large projects, the company chose simply to

Old rule:	Trust no one. Everyone is your enemy.
New rule:	Work as a team. Set shared goals and trust everyone's commitment to achieving those goals.
Old rule:	Point fingers. Fix blame. Never offer solutions that might expose you to additional liability.
New rule:	Get involved. Address problems quickly, while they are still small. Search for win/win solutions.
Old rule:	CYA. Document everything, early, often, and in elaborate detail before you act.
New rule:	Communicate openly with other team members. Trust that your words will not be used against you in the future. Document final decisions.

Figure 18.2 To make big gains, break the rules. Partnering breaks the rules of risk management, developed with the help of attorneys and insurance companies.

divide big projects into small work packages ranging from 7000 to 12,000 ft^2 each. These smaller packages could be easily executed by empowered individuals working in small teams.

Reynolds awards work packages to "partners" on a time and materials basis. These partners are measured by the timeliness of the work and their contribution to team-based performance goals. Long-term partnering relationships allow teams to move up learning curves toward improved schedule and cost performance. If a contractor or designer fails to perform, Reynolds no longer considers a claim or lawsuit, it simply does not invite that party back to work on future packages. Using this reengineered process, they have realized reductions in unit costs and in cycle times for planning, design, construction, and project management exceeding 50 percent.[8]

Reengineering the design firm. Partnering teaches participants that breaking the rules can pay big dividends. They learn that rewards await organizations bold enough to fundamentally redesign the way they deliver their services, systematically breaking the unspoken rules that limit their ability to change.

For example, design firms traditionally organized their work around the rule that information can be in only one place at a time. Information about a design was captured on paper and the paper was passed among team members. Work was organized sequentially, relying on review and coordination meetings to tie together work by various parties. New CADD and database technology makes it possible for all parties in the design process to access and use the same information simultaneously. This simultaneous design process holds the potential for facilitating design inputs, increasing coordination, and significantly shortening project schedules. Extended review periods, which were necessary under the old rule, disappear as all parties are able to review the status and detail of a design on an ongoing basis.

As these databases become more developed, design partners can also begin to "see" into each other's job files and project management control systems to access critical information on schedules and design requirements. This level of intimacy between project stakeholders relies heavily on long-term relationships and effective partnering to establish norms of trust and open communication.

These connections already exist in manufacturing. For its Saturn plant, General Motors reengineered its procurement process. GM allows suppliers to electronically "consult" its production schedule, formerly a closely held secret. In turn, suppliers are expected to use this information to plan their own production efforts to ensure that parts will show up when needed. The car manufacturer and suppliers took the process a step further by completely eliminating the purchase orders and invoices that used to drive the procurement process. No purchase orders are issued and no invoices are sent—suppliers build to suit GM's production system and are paid when parts arrive at the plant. Savings accrue to all parties in the form of reduced inventory costs, eliminated paperwork, and expedited payments.

This type of radical change is necessary for firms to remain competitive in the future. Clients that have paid the price, reengineering their organizations to improve quality, reduce costs, and reduce the time it takes to bring new products and services to market, are not about to be held hostage to design and construction processes that refuse to change. They are demanding that rules be broken. Partnering shows what can happen when they are.

Partnering for a Brighter Future

The design and construction industry may never again be able to "do it like it used to be done," but it does not have to repeat the sins of the past either. Trust and teamwork can be the basis of new, mutually beneficial working relationships among owners, contractors, and design firms.

The future looks considerably different for the remainder of the century than it did as recently as five years ago. The future is brighter, in no small measure due to changes that partnering has made possible.

References

1. *Partnering—A Concept for Success,* AGC of America, Washington, D.C., Sept. 1991.
2. Jon R. Katzenbach and Douglas K. Smith, *The Wisdom of Teams: Creating the High Performance Organization,* Harvard Business School Press, Boston, MA, 1993.
3. William H. Dadidow and Michael S. Malone, *The Virtual Corporation: Structuring*

and Revitalizing the Corporation for the 21st Century, HarperBusiness, New York, 1992.
4. Ray Stata, "Organizational Learning—The Key to Management Innovation," *Sloan Management Review,* Spring 1989, pp. 63–73.
5. Craig Barrett, "Organizational Learning in Practice," *McKinsey Quarterly,* Number 1, 1992, pp. 83–86.
6. David A. Garvin, "Building a Learning Organization," *Harvard Business Review,* July–Aug. 1993, pp. 78–91.
7. Michael Hammer and James Champy, *Reengineering the Corporation: A Manifesto for Business Revolution,* HarperCollins, New York, 1993.
8. Barry G. Lynch, R. J. Reynolds Tobacco Company, "Re-Inventing the Project Delivery Process Using Just-in-Time Project Delivery," *Proceedings of the AIA Conference: Cost, Time & Risk—Evaluating Project Delivery in the Face of Change,* March 1994.

Chapter 19

The Future of Partnering

Jerry Pitzrick

Now that project partnering has been shown to work, and its use is growing, what next? Jerry Pitzrick's contacts and brainstorming led him to envision a remarkable variety of possibilities, among them the following.

Construction companies commonly have tightly held power, linear authority hierarchies, and the mindset of doing things the way they have always been done. They may evolve toward more flexibility, empathy for others, and a willingness to try new things. If that happens, project teams will not accept "normal" performance, but constantly look for ways to improve schedule, cost, and customer satisfaction. Companies that work effectively with others may become the low-cost, high-quality producers.

Design firms may evolve toward "virtual" corporations. In a way they do this now by hiring contract workers to handle peak loads. This concept could be extended to involve the majority of the staff, working outside headquarters, either in smaller offices or at home. Now that most design is computer-aided design (CAD), people can have relatively inexpensive workstations at home and transfer files instantly by modem. In advertising this is being done today. Writers, art directors, print production people, and clients all work at separate locations and transfer files electronically. Using common software, having clearly defined roles, and a common understanding of the project objectives make these teams effective.

By now most of us in the A/E/C business have seen the range of success possible with partnering—from the very negative to the very positive. With that background, we are all wondering what is the future of partnering and how it will affect us. To get a clear picture of the future we need to take a critical look at the past.

The wide differences in success with partnering can be traced to a few items which influenced past outcomes, such as the level of senior management commitment, the effectiveness of the initial facilitation, the implementation of follow-up activities, and the existence of significant project challenges.

At a minimum, successful partnering has required the commitment of two out of three of the senior managers at these primary organizations: the owner, the design firm, and the general contractor. With this level of commitment, the third individual can usually be influenced by the others to become an effective contributor to the partnering effort. Senior-level commitment and involvement in the partnering process ensure that other shortcomings can be overcome. A lack of senior-level commitment and involvement dooms partnering.

Effective partnering facilitation is a balance between helping teams develop their personal interactive skills and building appropriate processes to make the team more productive at resolving challenging issues. In some cases, facilitators focus heavily on team development activities and improving interpersonal skills. This results in a concept of partnering as a "touchy feely," soft skills process which encourages people to work together but leaves them with minimal specific tools to accomplish the task at hand. When facilitators focus exclusively on identifying and solving project problems, people come up with solutions which help them on the project but have no effect on their long-term relationship. Only a combined emphasis on interpersonal skills and new problem-solving processes applied directly to the project can result in an effectively facilitated partnering workshop.

Another activity that influences the effectiveness of partnering is the implementation of follow-up activities after the initial workshop. Without follow-up, partnering turns into a single event with limited impact on the outcome of the project. Conversely, follow-up can reinforce new behaviors and help establish the agreed-upon processes. Effective follow-up can take many forms: Group problem-solving activities incorporated into regular project coordination meetings, facilitated sessions held on a quarterly basis, partnering meetings every six weeks, introductory meetings for new members of the project team, and social events can all be part of an effective follow-up effort.

The success of follow-up activities depends primarily on their regularity, the inclusion of all team members, a focus on team problem solving, and an effort to identify potential problems and deal with them before they affect the project. When these criteria are met, project partnering will be successful.

Partnering success may be affected by the presence of significant challenges on a project. Teams facing demanding schedule constraints, tight site conditions, environmental concerns, budget problems, or other extraordinary issues tend to experience extremes in the partnering experience—partnering may be considered either very successful or very unsuccessful. Such extraordinary elements amplify the effects of team efforts. Teams either accept the partnering concept wholeheartedly and learn to work through the challenges together, developing a strong sense of team unity in the process, or they fall back into old behaviors and resort to finger pointing and placing blame on individual members. Seldom do teams dealing with extraordinary issues end up with an ambivalent attitude toward partnering.

Management and team members at organizations experiencing this broad range of partnering effectiveness today are developing opinions about the effectiveness of partnering and their commitment to it in the future. This means that new teams on future projects may have diverse expectations and understandings of what partnering is and how well it can work for them. The obvious best case is for all team members to have had positive experiences and to know what they have to do to repeat them. At the other extreme, all team members have had negative experiences with partnering and agree not to pursue it on their next project. A combination of experiences with partnering will create the most significant challenges for future partnering success.

Changes for All Organizations

Currently, one of the primary objectives for partnering is dispute avoidance. This is a good beginning, but it only scratches the surface of the potential offered by partnering. Partnering provides an opportunity to develop truly high-performance project teams. To accomplish this will require a new level of organizational flexibility.

High-performance teams use processes that respond to the needs of the team. These processes are created from the collective input of all team members. Participating organizations need to allow their team members the flexibility to modify internal processes and systems when necessary to advance the project. This flexibility may result in role reversals in some organizations. For example, the people performing submittal reviews in an architect's office may need to have different responsibilities, with varying amounts of vendor contact, depending on the project. In the past, most office procedures and responsibilities were the same for all projects. That is, projects were supposed to be flexible to optimize the performance of the office. Now we want the office to be flexible to optimize the performance of the project.

As another example, owner organizations may have to reconsider how they administer the change process. Developing project-responsive flexibility on change processes may be especially difficult for large organizations. Systems for accumulating and summarizing project cost information will require a new level of flexibility to accept data coming out of the various change processes. People who now review all changes at exactly the same stage will be asked for their input at different times based on project requirements, and they may be asked not to participate at all at the time they are accustomed to reviewing changes. All organizations will be faced with similar challenges as A/E/C organizations build responsiveness to the needs of high-performance project teams.

Forms of contracts may also change as a result of partnering. A basic tenet of partnering is the equitable distribution of risk, although it is rare for all parties involved in a contract to talk about it in these terms. Partnering on private-sector projects readily allows for an equitable distribution of risk. On public projects, standard contract forms may be revised in non-project-specific partnering sessions with representatives of public agencies, architects, engineering firms, and contractors.

Contracts may also be simplified as a result of partnering. Traditional contracts focus on all the negative things that might happen on a project. This habit developed from accumulated negative experiences that people had on previous design and construction contracts. Contracts evolved to protect business entities from a repetition of bad experiences on past projects. As organizations involved in partnering accumulate positive experiences from partnering and earn the trust and respect of fellow participating organizations, contracts may be simplified to focus on defining the primary responsibilities of the participants. This would represent a fundamental change in how we define our relationships at the beginning of a project. If contracts define trusting relationships, project teams are more likely to respond with trusting actions. In this way partnering can lead the way toward a return of the days when contracts were something you looked at at the beginning of the job and then filed away and forgot. Simplified contracts will gradually evolve in the public sector as partnering grows and litigation therefore decreases.

The increased use of design-build contracts will create more opportunities for natural partnering relationships. However, the new forms of contract will not automatically result in partnering between designers and builders. Effective design-build teams integrate the typical design and construction processes into a single streamlined process which enhances the productivity of all participants.

Companies with design-build capabilities within one organization may have an advantage in integrating their processes for multiple projects, but will still need to remain open to incorporating the needs of the owner, subconsultants, and subcontractors to maximize the effectiveness of the project team.

Separate organizations coming together as a design-build team cannot rely on the contract form to make them effective. In fact, the assumption that the contract form will automatically lead to improvements in effectiveness may result in reduced team performance. People have also assumed that commitment to the contract by all parties will automatically result in improvement in team performance. Making this assumption will create problems because it creates a situation in which each person defines his or her role and approach to the work based on past experience. The lack of orientation to the current project may lead to the feeling that the work should be done one way because that worked on another project. Returning to the principles of partnering and jointly developing key processes and systems are necessary for real team effectiveness.

If we realize the full potential of partnering, we will see fundamental changes in many organizations. Companies will evolve from the current environment of tightly held power, linear authority hierarchies, and the mindset of doing things the way we have always done them to a more open environment in which flexibility, empathy for others, and a willingness to try new ways of doing things are the standard mode of operation. When project teams are created in companies with this attitude, the potential for major improvements will be excellent.

These teams will be the ones that will not accept "normal" performance. They will constantly be looking for ways to improve their project and will risk trying radically different ideas. These teams will also have the potential to make dramatic changes in schedule duration, cost of the work, and customer satisfaction. Because of this significant potential gain, an exciting aspect of partnering is that companies that do not embrace it could lose their competitiveness. Companies that work effectively with others will be the low-cost, high-quality producers.

There are many examples in industries other than construction of companies developing partnering relationships to improve their competitive position in the marketplace. U.S. automobile manufacturers have significantly decreased the number of suppliers they use and have developed stronger, more cooperative relationships with the ones they continue to use. (See Chap. 8.) They have also put new emphasis on strong relationships with their dealers. They are developing cooperative relationships with their employees. They realize that success depends on all parts of their system working together toward common goals of customer satisfaction, efficiency, and product innovation. All of these actions relate to similar situations the construction industry is starting to address with project partnering.

Another example that demonstrates partnering is the development of the Power PC microprocessor. Motorola, IBM, and Apple worked together to develop this new product. These companies have a cooperative relationship, but they also compete with each other in other areas. The construction industry can learn a lot about partnering relationships by looking at others outside our industry and learning from their experiences.

The Owners' View

Various federal agencies have made a strong commitment to project partnering. These include the U.S. Army Corps of Engineers, the General Services Administration (GSA), and the U.S. Navy. Commander Thomas Calhoun, Resident Officer in Charge of Construction at the Everett, Washington, Naval Station, has seen the benefits of partnering on their projects. As a result, his station is strongly considering requiring some form of partnering on all of their projects. The Navy has also started developing partnering relationships with design teams at the inception of a project. Other applications within the federal government include partnering with environmental agencies and the Occupational Safety and Health Administration (OSHA).

In most organizations, a quality effort works on standardizing processes to improve efficiency, eliminate errors, and improve customer satisfaction. Partnering, in contrast, looks at customized processes to improve project performance. Achieving the most effective balance between agency standardization and project customization is a significant challenge for the future.

One caution expressed by several owners is the potential for unscrupulous contractors to use partnering for their private advantage. This can be a significant concern for public agencies looking to standardizing partnering for all of their projects. Competitive bidding requirements minimize their flexibility in making contract awards. This could result in agencies reverting to more protective processes based on distrust and the need to protect one organization's position.

To minimize the potential of this happening, we all need to share partnering success stories and specific positive outcomes with other organizations. This will help encourage them to embrace partnering rather than reverting to old behaviors. Also, trade organizations may help by offering training sessions and round-table discussions on effective partnering practices. Peer pressure from other organizations will help eliminate the potential for companies to use partnering to manipulate outcomes for their sole benefit.

The GSA has implemented partnering on many projects and is now embracing it as an agency business strategy. Partnering began there as a tool for avoiding disputes with contractors and is now being used at the inception of projects. Partnering workshops are being held with building tenants at the conception stage of a project and with architectural and engineering firms at the start of design. To continue improving the partnering applications on their projects, the GSA is starting an internal training program on project partnering. This program focuses on developing a uniform approach to partnering for the agency and provides a venue for sharing successful practices and potential pitfalls. Jim Stewart, director of the Office of Design and Construction for the GSA, is a strong supporter of partnering and sees it as an effective tool to use together with reengineering and quality improvement to enhance the services provided by the GSA.

Quality improvement, reengineering, and partnering share a common emphasis on understanding the customer's needs, defining effective processes and using the input of diverse individuals. As partnering is standardized within the GSA, there is a concern that it not be institutionalized. The GSA realizes that the commitment to partnering is something people need to personally develop and demonstrate to others. Commitment cannot be mandated; that would just lead to people going through the steps but not really developing an effective partnering relationship.

To further support its partnering efforts in the future, the GSA is considering changing job descriptions to include greater emphasis on preventive activities for senior people and increasing the empowerment of project people to facilitate on-site decision making. The joint initiatives of reengineering, quality improvement, and partnering will also affect procurement practices. Diverse forms of procurement will be evaluated and, based on the needs of the projects, various options will be considered for each project.

Owners have found another benefit to partnering. When faced with a crisis project, they are able to call upon companies they have partnered with on previous projects, shortening the learning curve to respond to the new challenge. This has been a common practice in the private sector, where owners have used various forms of negotiated contracts to procure construction services. In the future, it will be easier for public agencies to respond to crisis projects similarly, because of the demonstrated success of their past working relationships.

A further improvement in relationships between owners and contractors is developing through non-project-specific partnering sessions. These sessions look at broader concerns which impact many projects and can be dealt with by changing procurement practices, owner/contractor contracts, owner/architect/engineer contracts, general contractor/subcontractor contracts, submittal processes, warranty processes, etc. (see, for instance, Chap. 17, on a new model subcontract). The input of diverse individuals creates opportunities that are not possible when single organizations make changes based only on input from their employees.

The Architects' and Engineers' View

Design firms are seeing the benefits of partnering during construction through more efficient submittal processes, fewer RFIs, and more cooperative problem solving. This has led to increased use of partnering from the inception of the design effort. Design-phase partnering will continue to grow and will include a broader range of participants. Representatives of regulatory agencies, facility occupants, public utilities, subconsultants, process equipment manufacturers, and local citizens' groups will all be considered as possible participants in partnering efforts. Providing these diverse groups with the opportunity for project input at the beginning of a project's design stage will create a stronger commitment to project success among all the individuals involved.

Other industries can provide examples of effective use of diverse inputs. Automobile manufacturers collect customer reactions to concept cars and use the information in new product development. Software developers frequently work with customers when developing new products. Even fast-food franchises work closely with customers and suppliers in developing new products. This broadening of input and welcoming of new partners into the project team will result in projects that can respond more effectively to diverse needs.

One of the more exciting applications of partnering may be in the development of design firms as "virtual corporations." Virtual corporations have been defined as organizations with few direct employees and relationships with a broad range of people and services. These virtual corporations can create various combinations of resources to respond to the unique needs of a project.

Design firms are presently using a limited version of this concept by hiring contract workers to deal with periods of high workloads.

This system could be expanded to include the majority of a design team, with many members working outside the firm's offices, either from smaller offices or from their own homes. This will decrease the need for large offices for companies and place workers in a more productive, worker-friendly work environment. Now that most design work is done on CAD systems, people can have relatively inexpensive workstations at home and transfer files by modem to other sites that need the information.

In the advertising industry, this is being done today. Writers, art directors, print production people, and clients all work at separate locations, transferring files to each other and building on others' work to develop a finished product. Using common software, having clearly defined roles, and a common understanding of the project objectives make these teams effective.

The typical construction project is much more complex than the development of a new sales brochure, but it too can be managed by the effective use of partnering meetings throughout the project. The initial role of partnering is to bring people together to develop key processes, clearly defined roles and responsibilities, and a common understanding of the project goals and objectives. Weekly or biweekly meetings can be held at a common location to review progress, work on problems, and identify future work assignments. These meetings would be critical, as they would be the only time people would meet face to face to work on the project. However, with effectively facilitated meetings responding to people's interpersonal needs and reinforcing the use of key processes, these teams could become more creative and productive than any group in today's typical organization. They will combine the benefits of a creative, flexible, and cost-effective home work environment with the benefits of a diverse team using efficient processes to achieve clearly defined common goals. This won't happen tomorrow, but using partnering within an organization can make it a reality.

The relationship between building designers and building-product manufacturers can also benefit from partnering. By identifying trends in the design industry and recurring problems, the two groups can jointly develop more effective products that respond to the needs identified by these discussions. Presently, new products are not readily accepted in the construction industry because of concerns about durability, compatibility with other materials, production quality control, production capacity, etc. If building-product development were a more interactive process, designers would have a clearer understanding of the answers to these questions through participation in defining the problems and the solutions. This would result in increased confidence in using a new product. Partnering sessions would help people identify specific problems which are appropriate for a diverse group to look at and establish processes to use their diverse knowledge to develop solutions.

The Contractors' View

Most partnering today takes place when contracts are in place and people are starting to work on a project. Partnering could start earlier. During the bidding process, a team could be assembled of a general contractor and several key subcontractors who would jointly develop a detailed plan on how they would build the project. This would require a commitment by all participants not to work with companies outside their team who are also pursuing the project. This prebid team would have a natural common goal of acquiring the project. They would be able to price their work more accurately because they would have a more detailed work plan with clearer roles for specific individuals. They could jointly identify unique approaches to the work based on their collective skills and their dependence on joint success.

Other side benefits would include a less chaotic bid day with less opportunity for bid errors, more accurate pricing of alternatives, and less stress in filling out complex bid forms under extreme time pressures.

Including many subcontractors as partners in the bid process may still be in the distant future. In the near term, however, more emphasis will be placed on recognizing subcontractors as critical partners during construction and placing equal value on their input in the partnering process. This will require more participation by subcontractors in partnering meetings and greater flexibility by general contractors to integrate this diverse input for the benefit of the project.

Currently, both general contractors and subcontractors have fallen into a trap where the general contractor makes unilateral decisions that affect everyone on the project. To move away from this process, all parties must develop trust and respect for each other. As people see the benefit of jointly developing solutions to small problems, they will be able to grow into applying this technique to larger problems. Future developments might include more direct communication among subcontractors, owners, and designers.

The construction industry in the United States has been criticized for a lack of commitment to research and development. Individual contractors have made improvements in techniques used by their companies, but little has been done to make significant breakthroughs in applying new technology in the industry. One reason for this has been the tremendous expense of funding major research projects.

Partnering could help with this dilemma. Competing general contractors could come together, identify a common problem, jointly develop a research team, fund the research, and then share the outcome. To be successful they would need to establish a joint commitment to the process and clearly defined roles and responsibilities. These are all issues that are commonly dealt with in developing a partnering relationship on a project basis.

The Subcontractors' and Suppliers' View

Specialty contractors' use of partnering can go beyond their role with a general contractor. Maintenance and operations contracts are big business for many subcontractors. Partnering can help companies move from providing a subcontracted service to being involved in a strategic alliance. An existing relationship can be expanded by obtaining a clearer understanding of each organization's needs and expectations and by using a more interactive process to develop solutions to problems. The starting point for this relationship is a subcontractor demonstrating its performance capability in its current relationship.

Once this is established, the next phase is the joint development of processes to respond to the needs of the owner. After the successful execution of these processes, the team can obtain a clear understanding of the business objectives of the owner and then, as a team, identify ways to improve the performance of the facility in meeting the owner's business objectives. The working sessions to identify these actions would be very open discussions, building on the diverse abilities of many individuals. Once again, this involves taking principles of partnering and applying them in a slightly different setting with wide participation.

Material suppliers may pursue partnering with designers. Additionally, they can develop relationships with research universities and government agencies, with the goal of developing new products. Many people will be anxious to share their ideas and support new product development if they are made aware of specific needs and if support is available.

On projects where owners and designers are not interested in developing a partnering relationship, anyone on the contracting team should be able to take the lead in developing partnering within the team of contractors. This team may not be able to influence overall project success to the same extent as a broader team, but it will be capable of improving the performance of all the construction companies.

Partnering may evolve to allow as small a group as all of the subcontractors on a hospital project doing work above the finished ceilings or in the walls. These contractors could focus on how to improve the performance of their collective work in these limited areas without affecting or being affected by others involved in the project. Of course, the broader the involvement the better, but partnering on even a limited basis can improve performance.

Labor Organizations' View

Labor unions and many contractors in the Minneapolis area have positive working relationships. One of the tools that has been used to create these relationships is labor/management joint safety initiatives on project sites. In these programs, union and company representatives jointly develop safety rules and consequences for the project. Project safety audits are routinely conducted by labor and management representatives. Unsafe conditions are corrected and unsafe actions are jointly addressed by union leaders and management.

This joint commitment to improving safety has led to discussions on how to be more proactive in eliminating safety concerns before they occur. The preventive actions identified include more detailed work planning, more clearly defined roles and responsibilities, and more effective training. All of these actions improve safety while also improving the quality of the work, job productivity, and working relationships. Just as in partnering between companies, labor/management partnering can significantly improve a project.

Ray Waldron, business manager for the Minneapolis Building and Construction Trades Council, has identified two areas for the entire construction industry to focus on for continuous improvement. One area involves establishing a process to capture the expertise of tradespeople and have it reflected on the design documents. The people who physically do the work have ideas for project improvements which never get incorporated into a project because they are not presented to the design team. A mechanism for worker involvement before the finalization of the design would allow these ideas to be reviewed and implemented from the onset of the project.

Another area Mr. Waldron suggests be focused on is the transfer of positive relationships to the next generation of union and contractor leadership. Without a special emphasis on transferring information, many of the positive factors in current relationships will be lost. Jointly developed archives of past successes and problems are one possibility that could serve as an effective learning tool for future representatives.

The Insurance Industry View

A basic tenet of partnering is the equitable distribution of risk while minimizing its impact on the project through effective team management. General liability and workers' compensation insurance is issued to general contractors and subcontractors, while errors-and-omissions insurance is provided to design firms. This separation of coverage requires that people look at risk management as a separate responsibility for each company, rather than focusing on the collective management of risk by the team. This focus on risk management does not affect the team until there is a significant problem. When a team is faced with a significant problem, however, it becomes most important for them to work together, but this is when insurance coverage can encourage finger pointing and the need to assign liability.

Wrap-up insurance programs which provide coverage to all participants on a project under one policy can help eliminate this finger pointing at such a critical time. The project team can then remain focused on solving the problem and may even be assisted by the insurance company at a time that is critical to project success. Today, insurers are pushing wrap-up aggressively on projects over $100 million. Insurance cost is often dramatically lower. Using wrap-up insurance on much smaller projects is a challenge for all parties.

Wrap-up programs could also be used to create incentives for the team based on total team performance. An example would be a project with no OSHA recordable accidents, which could have one-third of the saved premiums returned to the contractors and two-thirds retained by the owner. Other incentives could be developed focusing on team planning and team reward for performance.

DPIC Companies, Inc. (Monterey, California) provides errors-and-omissions coverage to many design professionals. The firm has seen sufficient benefit from partnering to pay for the cost of a partnering facilitator at the start of projects where it provides errors-and-omissions coverage. DPIC Senior Vice President, David Vermeulen, has seen partnering produce a positive change in attitudes resulting in a decreased number of claims and a lower average dollar value on the claims that are filed. As more people involved in the construction process see the benefits of partnering, the insurance industry will be there to provide more support for its successful implementation.

Facilitators' View

The current broad range of positive and negative partnering experiences will create new challenges for facilitators. Some team members will approach a project believing that partnering is essential to project success. Others will feel that it is a waste of time or that it had little impact on their last project.

To deal with this diversity of experience in an initial one- or two-day workshop could limit the productivity of the session. Facilitators or leaders of the companies involved in the partnering process will need to have small group meetings to talk about the team's diverse experiences and how the team members can jointly overcome any past negatives. When actions such as this are taken, the large group workshop will be more productive.

Partnering workshops have typically focused on two primary components: the development of interpersonal skills and relationships, and the identification of problems and techniques for groups to resolve them. Future workshops can continue to identify problems and develop solutions with little change in approach. Team development activities and personal profile systems are becoming repetitive and losing effectiveness with many groups. In groups where all participants have previously worked together, such activities may not be needed. With new teams, facilitators need to be creative and develop new exercises for team activity. Using worn-out exercises or skipping them completely will decrease the effectiveness of workshops. Partnering requires that facilitators assist in the development of the team's working relationships. To make this happen, facilitators need to take the lead in finding new ways of presenting necessary information.

Another challenge that facilitators face will be how to help an effective team move to a higher level of performance. On some projects, there are significant schedule demands, such as a $10 million project that must be completed within six months. Partnering has helped teams develop plans to successfully complete such projects on schedule. On other $10 million projects with a one-year schedule, teams have been less creative, but have successfully achieved the one-year completion. The pressure of a substantially shorter schedule has forced teams to a higher level of performance.

Facilitators can help all teams strive for this higher level of performance by creating exercises which help the team respond to higher challenges. Once solutions are developed by the team, the facilitator can ask the team to look at these unique solutions and see which can be applied to existing challenges. Such a process can produce results such as allowing a one-year project to be completed in eight months.

A final challenge for facilitators is eliminating the need for their involvement in follow-up activities and, eventually, even in initial workshops. Initial workshops can develop a structure for follow-up which does not require facilitators to be present. The challenge, then, is for the leaders of the participating organizations to assure that plans are implemented. The elimination of facilitators at initial sessions will arrive as our industry moves to partnering as the natural form of doing business.

The End of Partnering

Partnering will have achieved its ultimate success when the term "partnering" has died away and the principles of partnering remain as the usual approach to design and construction.

The challenge facing all of us in the industry today is to see that the principles remain. We must focus on effective interpersonal relationships, plan our work jointly, identify and solve problems collectively, and encourage creativity and innovation within each project team. These principles that we today call partnering can be expanded to how we work with our school systems and encourage students to continue their education in our industry, develop strategic alliances with owners which make them more competitive, as well as all stages in between.

We in the construction industry are at an exciting transition point that many other industries have faced and stumbled through to face the challenges of the twenty-first century. With the tools that partnering provides, we are in a position to control our destiny and move the construction process proactively toward ways of working together that assure joint success in the year 2000 and beyond.

Index

Accountability, 79–86, 106
Adopt-a-school program, 194–195
Advisors, attorneys as, 130–131
Advocates, attorneys as, 130–131
Agencies, federal (*see specific agencies, e.g.,* Army Corps of Engineers)
Air Force, 26–27
Al. Neyer, 2
Alternative dispute resolution (ADR), 14, 137–161
　baseball arbitration for, 155–156
　binding arbitration for, 155
　in contract documents, 140–143
　dispute review boards and, 145–147
　evolution of, 138–140
　mediation for, 147–152
　minitrials for, 152–155
　Project Neutral and, 157–161
　step negotiations for, 144–145
American Subcontractors Association (ASA), new model subcontract developed by (*see* New model subcontract)
Arbitration, 139–140
　baseball, 155–156
　binding, 155
　minitrials prior to, 154–155
Architects, 65–73
　advantages of partnering for, 71–73
　benefits of partnering for, 271, 290–292
　contract administration and, 68
　role of, 65–67, 70–71
　separation between constructors and, 67–70
　(*See also* Design firms)
Arizona Department of Transportation (ADOT), 10, 11

Army Corps of Engineers, 5, 288–290
　benefits of partnering for, 271
　first partnering project of, 272
　Fort Dix–McGuire Air Force Tertiary Wastewater Treatment Plant project and, 57–59
　TNT Thick Liquor Facility project and, 55–57
　(*See also* Public-sector partnering)
Arnold Engineering Development Center (AEDC), 21–22
Associated General Contractors of America (AGC):
　elements of partnering spelled out by, 232
　new model subcontract developed by (*see* New model subcontract)
Associated Specialty Contractors (ASC), new model subcontract developed by (*see* New model subcontract)
Attorneys, 13–14, 127–136
　as advisors and advocates, 130–131
　dispute resolution procedures complementing partnering and, 133–134
　drafting of contract language by, 132
　new model subcontract and, 262–266
　partnering as opportunity for more productive use of, 130–131
　traditional role of, 128–130
　uncertainties about role in partnering, 134–136

Bannes Consulting Group, 16–17
Baseball arbitration, 155–156

299

300 Index

BE&K Construction,
 management–employee partnering at, 15, 179–199
Beltsville Agricultural Research Center, 62
Benefits, family-centered, 190–192
Bid process, including subcontractors in, 292
Binding arbitration, 155
Bonneville Navigation Lock project, 32, 38
 evaluation form used for, 48–49
Budget control, 9
 accountability and mutual objectives for, 83–84
 partnering participants' evaluations of, 97–98

Celebration of successes, 190
Certification program, for excellence in construction, 117–118
Claims:
 under Contract Disputes Act, 54
 (*See also* Litigation)
Clients, private, 4–5
Close-out, 95
Coaches, middle managers as, 224–225
Collaborative problem solving, 270
Colorado steel-erection contractors, management–employee partnering and, 14–15, 163–176
Commercial factors, contractor choice and, 121–122
Commitment:
 to learning, 276–278
 of top management, 45–46, 63, 104, 284
Communication, of project scope and conditions, 249–251
Community service grants, 195
Conflict resolution (*see* Issue resolution)
Construction, preplanning of, 252
Construction companies, evolution of, 283
Construction Industry Institute, 5
Constructors:
 role of, 65–67
 separation between architects and, 67–70
 (*See also* Contractors)
Contract administration, architects and, 68
Contract Disputes Act, 54

Contractors:
 benefits of partnering for, 271, 292–293
 concerns of, 63
 partnering with, 16–17
 potential for abuse of partnering by, 288
 selection of (*see* Contractor selection; Strategic partnering)
 (*See also* Subcontractors)
Contractor selection, 116–117
 for Courtney Health Center project, 248–249
 (*See also* Strategic partnering)
Contracts:
 changes in, 286
 design/build, 286–287
 (*See also* New model subcontract; Partnering charter)
Convergence in choosing suppliers, 115
Corporate Development Services (CDS), strategic partnering projects and, 116–118
Costs:
 of arbitration, 140
 of litigation, 139
Courtney Health Center project, 245–255
 communication and analysis of scope and conditions and, 249–251
 construction preplanning and, 252
 contractor selection process for, 248–249
 marketing next job and, 254
 phases of contracting and, 251–252
 postconstruction phase of, 253–254
 project goals and, 247–248
CPM scheduling, 84
Craftworkers, 181–182
 control by, 181–182
 importance of, 182
 as product, 181
 training for, 193–194
Customers:
 Granite Rock's focus on, 215–222
 satisfaction of (*see* Customer satisfaction)
 surveys of, 217
 total quality management and [*see* Total quality management (TQM)]
Customer satisfaction:
 payment and, 217, 222
 total quality management and, 236

Decision making, benefits of partnering to, 9–10

Index

Delivery, contractor choice and, 121
Design/build contracts, 286–287
Design firms:
  evolution of, 283
  reengineering, 280–281
  (*See also* Architects)
Dignity, of employees, 185–187
Dispute resolution (*see* Issue resolution)
Dispute review boards (DRBs), 145–147
Donald B. Murphy Contractors (DBM), 31–51
  Bonneville Navigation Lock project and, 32, 38, 48–49
DuPont Corporation, strategic partnering at, 113, 115–118, 123
Duties, incorporating into contracts, 140–141

Empathy, 35–37
Employee empowerment:
  accountability and mutual objectives for, 81, 83
  decision making and, 9–10
  of employees, 222, 224
Employees:
  celebration of successes and, 190
  community involvement of, 194–195
  dignity of, 185–187
  empowerment of (*see* Employee empowerment)
  family-centered benefits and, 190–192
  hiring of, 222
  informed, 192–194
  partnering with (*see* Management–employee partnering)
  partnering with [*see* Management–employee partnering; Total quality management (TQM)]
  recognizing worth to company, 183
  skilled, shortage of, 181–182
  (*See also* Craftworkers)
Empowerment:
  of customers, 217, 222
  of employees (*see* Employee empowerment)
Engineers:
  benefits of partnering for, 290–292
  (*See also* Army Corps of Engineers)
Equity, 39–40
Evaluation form, 48–49

Expectations:
  importance of, 104
  positive, 11–12

Facilitators, 61, 295–296
  quality of, 47
Family-centered benefits, 190–192
Far East, management–employee partnering in, 205–208
Federal agencies (*see specific agencies, e.g.,* Army Corps of Engineers)
Field implementation, 94–95
Fixed-price partnering (*see* Public-sector partnering)
Flexibility, project-responsive, 286
Follow-up, 105, 284–285
  of Courtney Health Center project, 253–254
Ford Motor Company, strategic partnering at, 113, 115–116, 120–123
Foreign firms (*see* Global partnering)
Fort Dix–McGuire Air Force Tertiary Wastewater Treatment Plant project, 57–59
FRU CON, 22–26

General Motors, 281
General Services Administration (GSA), 288–290, 289
Global partnering:
  in Far East, 205–208
  international practice and, 203–204
  in North Africa and Middle East, 204–205
  at Parsons Brinckerhoff, 201–210
  in Turkey, 208
Goals:
  company, employees' personal goals and, 183–184
  employee identification with, 182–183
  measurement and monitoring of, 105–106
  setting (*see* Goal setting)
  "stretch," 226–227
Goal setting, 270
  cooperative, 247–248
Government agencies (*see specific agencies, e.g.,* Army Corps of Engineers)
Graff Flooring, 17
Granite Rock Company, 211–229
  customer focus at, 215–222

Index

Granite Rock Company, (Cont.)
  employee empowerment at, 222, 224
  employee training at, 225–226
  hiring at, 222
  as Malcolm Baldrige National Quality Award winner, 211–215
  before partnering, 215
  performance measurement and goals at, 226–227
  role of middle managers at, 224–225
Grants, community service, 195

Honor, 42–45

Identification, of employees with company, 182–183
Informational meetings, 192
Informed employees, 192–194
Innovation, 47, 49–50
  accountability and mutual objectives for, 85–86
Insurance industry, benefits of partnering for, 294–295
Intel Corporation, strategic partnering at, 113, 115–116, 118–120, 123
International partnering (see Global partnering)
Issue resolution:
  adherence to process for, 106
  attorney's role in, 133–134
  importance of early resolution and, 10
  partnering participants' evaluations of, 101–102
  shared stakes and, 37–39
  TNT Thick Liquor Facility project and, 55–57
Issue resolution process phase of preproject workshop, 92, 94

Jargon, total quality management, avoiding, 228
Job-site personnel, inclusion of, 105

Kaizen, 272

Labor unions, benefits of partnering for, 294

Large Rocket Test Facility (LRTF) project, 22
Laughlin Water Reclamation Facility project, 76–87
Lawyers (see Attorneys)
Leadership, 33–35
Learning, commitment to, 276–278
Learning organizations, 278
Litigation, 13, 54, 128–130
  causes of, 139

McDevitt Street Bovis, 16
  total quality management at, 231–244
Malcolm Baldrige National Quality Award, Granite Rock Company as winner of, 211–215
Management:
  coach role for middle managers and, 224–225
  commitment of, 45–46, 63, 104, 284
  for total quality management, 237–238
Management–employee partnering, 14–16
  at BE&K Construction, 15, 179–199
  Colorado steel-erection contractors and, 14–15, 163–176
  employee dignity and, 185–187
  employee involvement and, 194–195
  family-centered benefits and, 190–192
  informing employees and, 192–194
  management–employee partnering at, 15–16
  at Parsons Brinckerhoff, 15–16, 201–210
  reasons for, 179–182
  safety and, 187–190
  success factors for, 183–184
  theory of, 182–184
Martin K. Eby Construction Co., 8
Marvin M. Black Excellence in Partnering Awards, 11
Material suppliers, benefits of partnering for, 293–294
MCI Constructors:
  Fort Dix–McGuire Air Force Tertiary Wastewater Treatment Plant project and, 57–59
  Metropolitan Washington Airports Authority and, 62
  switch to partnering, 53–63
  TNT Thick Liquor Facility project and, 55–57

Mediation, 147–152
Meetings, 46–47
  informational, 192
Metropolitan Washington Airports Authority, 62
Middle East, management–employee partnering in, 204–205
Minitrials, 152–155
  prearbitration, 154–155
Multiproject strategic partnering, 12–13
Mutual objectives, 79–86
Myers Briggs Test Indicator (MBTI), 23

Natural Resources Agencies Building project, 72
Navy, 288–290
Negotiations, step, 144–145
New model subcontract, 257–266
  legal consultation for, 262–266
  writing of, 260–262
North Africa, management–employee partnering in, 204–205

Objectives:
  mutual, 79–86
  (*See also* Goal setting; Goals)
Occupational Safety and Health Administration (OSHA), steel erection safety and, 163–176
Oliver Lock and Dam project, 22–26
Organizational change, 287
Organizational structure, teams and, 275–276
Organizations:
  effect of partnering on, 29
  learning, 278
Owners:
  benefits of partnering for, 288–290
  uncertainties about legal issues involving, 135–136

Parsons Brinckerhoff, management–employee partnering, 15–16, 201–210
Participants in partnering (*see* Partnering participants; *specific participants*)
Partnering:
  benefits of, 271–272
  as bridge to total quality management, 272–273

Partnering: (*Cont.*)
  definition of, 2, 270–271
  future of, 17–18, 283–297
  as learning process, 277–278
  limitations and disadvantages of, 63
  non-project-specific, 290
  participants in, 11
  as strategic initiative, 104
  trends in, 106, 108–109
  variations in success of, 75
Partnering charters, 47, 128, 270
  dispute management in, 140–143
  risk shifting and, 132
  uncertainties about legal status of, 134–135
  (*See also* New model subcontract)
Partnering facilitation, 284
Partnering participants, 95–104
  on budget control, 97–98
  on completed projects, 102–103
  on issue resolution, 101–102
  overall evaluation of partnering by, 103–104
  on quality, 99–100
  on safety, 98–99
  on scheduling, 96–97
  on working relationships, 100–101
  (*See also specific participants*)
Partnering process, 89–109
  beginning of, 104–105
  close-out and, 95
  critical success factors for, 104–106
  field implementation and, 94–95
  participants' views of, 95–104
  preparation phase of, 91
  preproject partnering workshop and, 91–94
  sustaining, 105
Payment, customer empowerment and, 217, 222
Performance measurement, 226–227
Plan-Do-Check-Act (PDCA) process, 236–237
Planning:
  before construction phase, 252
  total quality management and, 239–244
Positive expectations, 11–12
Preparation, total quality management and, 241–242
Preparation phase of preproject workshop, 91
Preproject workshops, 91–94, 270–271
Private clients, 4–5

Problem solving:
  accountability and mutual objectives for, 85–86
  collaborative, 270
Problem-solving phase, of preproject workshop, 94
Project charter phase, of preproject workshop, 92
Project delivery process, teams and, 275
Project Neutral, 14, 157–161
  applications of, 160–161
Project schedules (*see* Schedules)
Project success, partnering participants' evaluations of, 102–103
Project teams (*see* Teams)
Public projects, 5–6
Public-sector partnering, 19–29
  Air Force and, 26–27
  future of, 29
  introduction of, 21–22
  Oliver Lock and Dam project and, 22–26
  organizational change brought on by, 29
  problems encountered in, 25–26
  reasons for success of, 27–29
  (*See also* Army Corps of Engineers)

Quality:
  accountability and mutual objectives for, 83
  contractor choice and, 120–121
  partnering participants' evaluations of, 99–100
  strategic partnering for (*see* Strategic partnering)
  [*See also* Total quality management (TQM)]
Quality planning, 239–244

R. J. Reynolds Tobacco Co., 269, 279–280
Recognition of employees' worth to company, 183
Reinforcement, 105
Relationship building, total quality management and, 242
Resolving issues (*see* Issue resolution)
Resource commitment, 105
Respect for employees, total quality management and, 238–239
Responsibilities, incorporating into contracts, 140–141
Rights, incorporating into contracts, 140–141
Risk sharing, incorporating into contracts, 141
Risk shifting, 132

Safety, 8–9
  accountability and mutual objectives for, 80–81
  Colorado steel-erection contractors and, 14–15, 163–176
  management–employee partnering and, 187–190
  partnering participants' evaluations of, 98–99
  reporting of near misses and, 188, 190
  site safety committees and, 188
  special emphasis programs and, 188
  standards for, 187–188
Saturn plant, 281
Schedules, 8
  accountability and mutual objectives for, 84
  partnering participants' evaluations of, 96–97
Senior management, commitment of, 45–46, 63, 104, 284
Site safety committees, 188
Skill improvement:
  total quality management and, 242
  (*See also* Training)
Southeastern University, 62
Southwest Airlines, 222
Special project groups, 193
Step negotiations, 144–145
Strategic partnering, 113–125
  at DuPont, 113, 115–118, 123
  at Ford, 113, 115–116, 120–123
  information resources on, 125
  at Intel, 113, 115–116, 118–120, 123
  success factors for, 114–116
Subcontractors:
  benefits of partnering for, 293–294
  including in bid process, 292
  partnering with, 16–17
  uncertainties about legal issues involving, 135
Successes, celebration of, 190
Success factors:
  for management–employee partnering, 183–184
  for partnering, 104–106

Success factors: (Cont.)
  for strategic partnering, 114–116
  for total quality management programs, 227–229
Sundt Corporation, 4
Supplier Continuous Quality Improvement (SCQI) program (Intel), 118–120, 123
Supplier Quality Evaluation (SQE) program (Ford), 120–123
Suppliers:
  benefits of partnering for, 293–294
  (See also Contractors; Subcontractors)
Surveys of customers, 217
Sylvan Industrial Piping, 123

Team report card phase of preproject workshop, 92
Teams, 274–276, 287
  attorneys as members of (see Attorneys)
  building, 45, 128, 270
  measurement and monitoring of goals of, 105–106
  organizational structure and, 275–276
  responding to needs of, 285–286
  transformation of project delivery process by, 275
Technical factors, contractor choice and, 121
Time commitment, 105
TNT Thick Liquor Facility project, 55–57
Top management, commitment of, 45–46, 63, 104, 284
Total quality management (TQM), 16, 76
  communication necessary for, 85–86
  customer focus and, 227–228

Total quality management (TQM), (Cont.)
  customer satisfaction and, 236
  at McDevitt Street Bovis, 231–244
  Malcolm Baldrige National Quality Award and, 211–215
  partnering as bridge to, 272–273
  Plan-Do-Check-Act process for, 236–237
  planning and, 239–243
  reasons for, 233–234
  respect for people and, 238–239
  success factors for, 227–229
Training, 225–226
  about company, 192–193
  in crafts, 193–194
Trust, 40–42
  building, 77–79
  total quality management programs and, 228–229
Turkey, management–employee partnering in, 208

Unions, benefits of partnering for, 294
U. S. agencies (see specific agencies, e.g., Army Corps of Engineers)

Visibility of partnering, 106

Wages, management–employee partnering and, 180–181
Whiting-Turner, 4–5
Workers (see Craftworkers; Employees)
Working relationships, partnering participants' evaluations of, 100–101
Workshops, preproject, 91–94, 270–271